HUXLEY IN AMERICA

HUXLEY IN AMERICA

MICHAEL COLLIE

ACADEMICA PRESS

BETHESDA—PALO ALTO - DUBLIN

Library of Congress Cataloguing-in-Publication Data

Collie, Michael.
Huxley in America : T.H. Huxley , American science, university education and scientific propaganda, 1870-1880 / Michael Collie.
p. cm.

Includes bibliographical references and index.

ISBN 978-1-936320-06-6

1. Huxley, Thomas Henry, 1825-1895—Travel—United States. 2. Huxley, Thomas Henry, 1825-1895—Influence. 3. Naturalists—Great Britain—Biography. 4. United States—Civilization—1865-1918. 5. Science—United States—History—19th century. 6. Science—United States—Philosophy—History—19th century. 7. United States—Civilization—British influences. I. Title.
QH31.H9C75 2011
508.092—dc22

2011001274

Academica Press, LLC
Box 60728
Cambridge Station
Palo Alto, CA. 94306

Website: www.academicapress.com

to order: 650-329-0685

For Joanne Collie

I cast off my costly garments
Donned the working clothes you see
And the harp that was my music
Hung upon a willow tree

CONTENTS

ACKNOWLEDGMENTS

Some years have passed since I visited the libraries listed towards the end of the present book, while what became a global survey of unpublished Huxley material was being conducted, and I am happy to record my gratitude for the invaluable assistance I received at that time from librarians and archivists. Based as it is on actual visits to libraries with holdings of Huxley manuscripts and printed books, the present monograph would have been impossible without their generous help.

More recently I have been greatly assisted by Joanne Collie, Leslie Howsam and Thomas Knowles."

PREFACE

This book is intended as a modest contribution to the understanding of an extremely influential, nineteenth-century British scientist, Thomas Henry Huxley, who distinguished himself by devoting his best energies to the idea that science should be central to everyday life, and particularly to the education of the young. The narrative and discussion that follow mostly concern Huxley's 1876 visit to North America, because, in addition to the fact that no good account of his visit currently exists, it is believed his experiences in that year helped define the direction his life would take from then on. He would be an advocate for science but not a research scientist.

Within the modern academic discipline of the history of science there are those who quite properly believe that science, especially with regard to its more important findings and results, can be thought about, even should be thought about, quite separately from the people who happen to conduct it. In this context the science is much more important than the scientist. A different approach has been adopted here. In the present book, a complementary line of enquiry has been adopted, an approach from which the character, motivation, rational and irrational preferences, and deeply imbedded habits of mind of the subject are not excluded. This is because the ways in which Huxley frequently thought and behaved are so unusual as to cry out for discussion and explanation. He was not a normal human being. He was an exceptional human being who became President of the Royal Society of London, but one who, nonetheless, paid a price, psychologically, for his success. It is this that the present book begins to investigate.

INTRODUCTION

The intention in this new book about the British nineteenth-century scientist, T.H. Huxley, is to provide an accurate account of his visit to the United States in 1876, a visit which included a week's work with the distinguished palaeontologist, O.M. Marsh at Yale College, a journey west as far as Buffalo and Niagara, a longer journey to the South to visit family, the delivery of the inaugural address at the opening of the Johns Hopkins University and finally a short stay in New York City where he gave three lectures on evolution. Between these major events were other visits, including brief stops in Boston, Cambridge, Philadelphia and Washington. This was one of the pivotal years in Huxley's long, complex and impressive career.

Closely related to the actual visit were other activities, commitments and research enquiries, notably Huxley's involvement in the founding of the International Scientific Series, his long-term relationship with the New York publishing firm of D. Appleton & Co., and his contributions to the Humboldt Library. Each of these will be described and analysed in all necessary detail.

Intellectually, the visit resulted in a significant advance in Huxley's thinking about theories of evolution. At Yale, and perhaps elsewhere, he learned things he had not previously known or understood. As will be seen, he approached the concept of evolution as a comparative anatomist whose experience had taught him that the bodily structure of most animals, including man, had remained constant for long periods of time. For him to think otherwise would require compelling evidence. Perhaps he would acquire such evidence when he crossed the Atlantic. All in all, his boldly conceived American visit was an immense success, coming, as it did, at a perilously crucial point in his career. In 1876 he was poised between being a still-active research scientist who could see that the time had come for him to extend the boundaries of his work beyond the merely descriptive or, on the other hand, concentrating his best energies on education and the extension of other people's awareness of the importance of science.

From the time that he was back in London from Australia and had resigned from the Navy, he had been an aggressively competitive, career-conscious scientist. He very much wanted to be first among equals. More than a few of his colleagues had similar ambitions, which they satisfied in their various ways. Among his scientific peer-group in London, there were colleagues who travelled mostly, or even exclusively, on the Continent in order to know about and understand what European scientists were doing, others who looked to the British colonies for the acquisition of new knowledge and new insights in their particular research fields, and many others who never travelled at all, remaining distinctly insular and sometimes chauvinistic in their attitudes though nonetheless active in their chosen fields of enquiry. In the first category was Huxley's boss, Roderick Murchison, whose familiarity with European scientists of that era was unparalleled. His French was fluent, his German and Italian adequate, and he had made a point of travelling each summer to various parts of Europe, including Scandinavia. In the second category were people like J. B. Hooker and Hugh Falconer, much of whose research was conducted in India. Both men, as a result, came to know a great deal that was previously unknown, indeed was still inaccessible to most Europeans. If the third category can be discreetly passed over, staying in the country where you happened to find yourself, despite obvious limitations, was perfectly natural and normal for most people; it will be enough to say that Huxley was one of the few who learnt by experience that contact with American scientists might be extremely beneficial, both intellectually and in terms of career advancement, as indeed it turned out to be. He did not, however, conduct his own research in North America, so one of the themes of the present book will be his habitual dependence on other people's fieldwork, publications, syntheses and communications. Was he an original thinker or something of an academic parasite?

But first we must ask: who was the Thomas Henry Huxley who will be the subject of this book? The short answer is that he was a prominent, nineteenth-century British scientist who had trained as a medical student at the Charing Cross Hospital in London, was subsequently ruled intellectually by the knowledge of anatomy he had acquired at medical school, eventually wrote more than three hundred scientific papers and articles, as well as between thirty and forty books, and succeeded in imposing himself on the burgeoning Metropolitan science scene by force of character and an unfailing belief in the social and educational importance of a scientific education. There never existed a more energetic, whole-hearted and knowledgeable ambassador for science. By 1876, when he went to America, he was already over fifty years old, an impressive figure physically who would immediately attract attention whenever he entered a room or mingled with people who had not previously met him. A student who attended some of his lectures during the eighteen seventies described him as "a tall old man with white hair and beard and heavy eyebrows, his face ruddy with a glow that I think must have been due to excitement"[1].

His early years, which are not at all part of the present narrative, are reasonably well known, so only the briefest of summaries is needed here. The salient events of his life prior to the American expedition were an education as a medical student at the Charing Cross Hospital in London, which carried with it into the future a permanent predisposition to regard anything anatomical as fixed, unchanging and unchangeable, skeletons being more interesting than flesh and blood human beings or animals; his long voyage on HMS Rattlesnake to Australia, which gave rise to the various papers he submitted to the Royal Society of London—substantial, original works that seemed to launch him into the world

[1] University of Edinburgh Library: Dk.8.42. The writer of the note is not identified.

of research enquiry; his determined, full-hearted involvement in Metropolitan science, and his election to the Royal Society and early membership of other London societies and associations. He had become a Londoner with all the advantages and disadvantages that entailed. The die was cast at this point. His thinking would remain institutional for the rest of his life and London, for him, was the institutional centre of the universe.

Between his return from Australia and his double appointment in 1854 as lecturer at the School of Mines and palaeontologist at the Geological Survey, appointments, which were quickly followed by his becoming a Fullerian Professor of Physiology at the Royal Institution in 1855, he demonstrated immense fortitude in resolutely sticking to his chosen career path, being vindicated eventually, though not by 1854, since at that point he knew next to nothing about the burgeoning new science of palaeontology, while at the same time he had to teach himself, more or less from scratch as he admitted (or boasted), what would become, for him, an effective lecturing technique on which he could depend. It was, however, undoubtedly a success story. Starting with very little, he succeeded in establishing himself at the very heart of scientific Britain.

Huxley's early career was by no means simply the story of a determined, strong at heart, highly intelligent, apparently resilient individual working steadily towards his chosen goal, although these were qualities on which he was always able to depend. There was a long way to go, and many obstacles to overcome, before he would be elected President of the Royal Society of London. In the interim, the contradictions of character, which may indeed have contributed in a number of ways to his career success, somewhat ensured that life would be far from easy. On the one hand, he was often seen as a man of considerable charm, someone any person would wish to know. "He really is a most agreeable man", Leslie Stephen would say, writing to Minnie Thackeray, "considering that he is a scientific swell".[2] He was often conciliatory, as when he made an effort to get on with Louis Agassiz, "hoping", he said, "he will henceforth be marked by you neither among the cantankerous nor even the bumptious".[3] Later in life, he would try to establish a friendly relationship with the novelist, Henry James, though without understanding why the desire was not reciprocal. While in North America he would form friendships that lasted a lifetime with Othniel Marsh, Spencer Baird, William Appleton and others. Many of his English friendships were also life-long.

On the other hand, he sometimes even boasted about being difficult to get on with, competitive at the expense of others, and obstinate in personal conflict. For example, he would tell his geologist friend, Archibald Geikie, in a letter from Edinburgh, that he had to sustain "fighting enough to prevent me from going blue mouldy for want of a bating".[4] Earlier, he had written that "fighting and scratching to keep your place in the crowd exclude almost all other thought".[5] Nor did he ever gain complete control of his aggressive nature, for in 1873 he is writing to his friend Knowles in the same uncompromising terms: "I fought the parsons at Aberdeen and got beaten but the press was on my side & their victory will not serve them long".[6] This conflict of character between the conciliatory and the hostile, where the conciliatory could be genuine, as in his belated attempts to recover the grounds of friendship with Herbert Spencer and where the hostility could be life-long, as in his undisguised, pathological hatred of Roderick Murchison, has never been very much written about, possibly because the emphasis has so often, rightly, been on his undoubted achievements and his, in some respects, unique philosophical standpoint.

---

[2] ALS: Perkins Library, Duke University

[3] ALS: 11 July 1859, Houghton Library, Harvard University. Louis Agassiz and his family had moved from Switzerland to Harvard many years before this date.

[4] ALS: 19 October 1862, Edinburgh University Library

[5] ALS: 6 November 1851, Imperial College, London

[6] ALS: 25 April 1873, American Philosophical Society, Philadelphia

There are in any case other factors that must be taken into account by anyone wishing to understand Huxley's complex psychological make-up and the apparent anomalies of character. One of these was hard work, what nowadays some people call the Victorian work ethic, often seen as a self-justifying activity, or rather that dangerous capacity for over-work that sometimes can befuddle any individual who, sincerely believing he is doing his best, in his own terms, loses touch with the plain realities of existence as observed by others. This was certainly Huxley's position. He had worked and worked to achieve social status to the point that work had become an end in itself. If you were working flat out, what more could you do? At the same time, it is reasonable to keep in mind, when trying to understand his motivation, that one of the harsh realities of his private existence had been the death of his young son, which he once described as "a knock down blow" from which he only recovered, he said, "by patience and hard work".[7]

By the mid-eighteen sixties, having started with very little and having worked energetically to establish himself at the heart of scientific Britain, it became obvious he had been driving himself too hard. Members of Huxley's family could see this was the case. "Hal is very tired", Nettie (his daughter) told John Tyndall in March 1868; "For three years I have been urging him to give up the College of Surgeons lectures—which coming as they do just after the course at Jermyn Street are too much for him".[8] The School of Mines and the administrative headquarters of the Geological Survey were at that time in Jermyn Street, which is just a few minutes walk from Piccadilly Circus in the heart of London. In April 1870 he explained to George Gordon: "I am so terribly tired & used up with a very hard winter's work that I am recommended to get a week or ten days of complete change away from the fossils & all their works—and so I am going to run away to the Rhine country & have a tramp there with my old friend

---

[7] Letter to Charles Lyell
[8] ALS: 12 March 1868, Royal Institution, London, #2918

Dr. Hooker".[9] He had by then been working with the Rev. Dr. George Gordon for several years on the description and analysis of the fossil remains of extinct reptiles at that time being discovered along the shores of the Moray Firth in North-Eastern Scotland.[10] Important and fascinating though the work had been, it proved too much for him. A year later the situation became more serious. "I have been so ill", he told Archibald Geikie in February 1871.[11] This is the as yet not fully diagnosed illness from which he suffered during the five years preceding his trip to North America. In July of that year he again wrote to George Gordon: "I am going to give myself a serious holiday this year—and to that end have taken a house in St. Andrews for the months of August and September".[12] He used the time, in part, to judge whether the university would be the right place for his son. Just two years later, Charles Darwin and a group of other friends who very much cared about Huxley's well-being raised a large sum of money—over £2000—to enable their friend and colleague to take a break from his unremitting work schedule, a schedule which had to do both with a consuming interest in and devotion to scientific research, as he understood it during the pre-America years, and his desire to support his wife and family in the manner he judged appropriate. Possibly those who knew Huxley best, perceiving that his desire for a substantial income had somewhat displaced the desire for a significant career as a research scientist, hatched this generous plan to help him return to science. If so, the plan did not work. He was genuinely ill, though with a condition his doctor was unable to diagnose with full authority. "For the first time in my life I have been beaten, said Huxley. "I mean morally beaten. I have for months been without energy and without hope & haunted by the constant presence of hypochondrial apprehensions, which my reason told me were absurd, but which I could not get

---

[9] ALS: to George Gordon, 14 April 1870. Huxley and George Gordon had been in touch with each other since 1858, having worked together on the fossil remains of extinct reptiles being found in the north-east of Scotland.

[10] See Collie *Huxley at Work* (Basingstoke: Macmillan, 1991)

[11] ALS: University of Edinburgh Library. Archibald Geikie was employed by the Geological Survey of Great Britain of which he subsequently became Director.

[12] ALS: to George Gordon, 22 July 1871; DDW71/870/3

rid of. No, I was breaking down".[13] He later explained this to Gordon: "I am sorry that my hands are still full—all the fuller for a year and a half's illness, which has kept me in a woeful condition of incapacity".[14] There is a letter to an unnamed correspondent in the Houghton Library with a probable date of 1 April 1873 saying he believes he is recovering, "provided always", he said in a different letter to Knowles written at about the same time, "that I drink no wine, eat as little as nature can do upon, write no articles, and give my time to no *Metaphysical Discussions*".[15] In other words there is no doubt whatsoever that he had been seriously ill, that the long illness had affected his work and that he pretended it had not.

The high level of exhausting professional and scholarly activity can be seen in Huxley's work as a translator, in the writing of major articles at the rate of five or six a year, and particularly in the publication of the books listed in Appendix One of the present study. Wanting to have a finger in every pie, while earnestly supporting the family he loved in the manner he judged appropriate—a separate bedroom for each of his children, he explained—he had simply been doing much too much, perhaps with more of an emphasis on quantity than quality, in as far as many of the articles were just straightforwardly descriptive, and perhaps written in an aggressive spirit that, ultimately, was psychologically damaging, or so it is reasonable to ask. Leonard Huxley quoted from a letter in which Huxley said: "I have amused myself by spending ten days or so in bed. I had an unaccountable prostration of strength which they called influenza but which, I believe, was nothing but some obstruction to the liver".[16] Interestingly, he ends this letter to Darwin, dated May 1872, in the following way: "I have come to the conviction, however, that steady work hurts nobody, the real destroyer of hard-working men being not their work, but dinners, late hours, and

[13] ALS: 24 April 1873, Cambridge University Library
[14] ALS: to George Gordon 28 July 1874; DDW/871/18
[15] Letter-text reproduced by Leonard Huxley, Volume One, p. 206
[16] ibid.

the universal humbug and excitement of society. I mean to get out of all that and keep out of it".[17] Crossing the Atlantic was one of the ways in which he attempted to achieve this goal.

At all events, readers new to Huxley need to know not only about the great achievements but also about the mental turmoil and fatigue he experienced during the early eighteen seventies and, with this in mind, perhaps be better able to assess the significance of his American travels which began, of course, with a recuperative sea-voyage that took him completely away from the scientific in-fighting of the English metropolis. The previous attempts to get him to relax had failed. He had been to Egypt. He had walked along the Rhine. No doubt other steps had been taken to restore him to full health. Would an increased knowledge of the American way of life be helpful to him or not? Would an extension of experience in North America settle his nerves, allowing him to view his own life and his own career in a less troubled manner, with a more accurate type of self-knowledge? Clearly one of the several elements in the planning of the trip was therapeutic. He needed a break. He needed to get away. Would he benefit? And if so, in what ways?

Because of the complex manner in which Huxley's career developed between his appointments in 1854 and his journey to North America fifteen years later, there must undoubtedly be not one, but several valid ways by which to locate him socially and intellectually, especially if the purpose is to understand what mattered to him, what motivated him, what drove him to work unremittingly. Consequently, if one approach is adopted here for the purposes of the present book, it is by no means to say that an entirely different approach might not also be useful. Indeed, some critics have recently discussed Huxley without even mentioning what he wrote and published, or that he wrote and published as much as he did.

---

[17] ibid.

Here, however, for the immediate purpose, his little book called *Evidence as to Man's Place in Nature*[18] first published in London by Williams & Norgate in 1863, will be used to help identify and locate the problems this medically-trained comparative anatomist faced as he tried to understand and come to terms with the implications of Darwin's *The Origin of Species* published in 1859. It is well known that Huxley would, and did, support Darwin vociferously whenever the concept of evolution was attacked by clergymen, since Huxley's dislike, even hatred for the Church of England was lifelong, and because he was endlessly determined to demonstrate that science was completely independent of religion, but at the same time, at a deeper level, coming to a scientific understanding of how an organism could be structurally modified over time remained a problem for him, a problem fairly clearly stated in *Evidence as to Man's Place in Nature* and then, as will be seen, to some extent resolved in America during the visit described in this book. *Evidence as to Man's Place in Nature* was the statement of the problem; meeting scientists like O.M. Marsh at Yale would significantly help him towards a possible resolution of it. In 1863, he was predisposed to believe that the anatomical structure of each and every creature on earth, including man, was a fixed, never changing entity that could be demonstrated as such in the lecture room, which is what he did. Would he have to change his views?

The book starts with an "Advertisement to the Reader", dated January 1863 where he says "the greater part of the following Essay has already been published in the form of Oral Discourses, addressed to widely different audiences, during the past three years".[19] This is typical. Huxley was already becoming the Grand Master in the recycling business. The term "oral discourses" is problematic. He is referring to two sets of lectures, the first to "working men" in London in 1860, and the second at the Philosophical Institution in Edinburgh in

---

[18] *Evidence as to Man's Place in Nature* (London: Williams and Norgate, 1863)
[19] Ibid. In the advertisement he says he hopes he has not committed the error "into which working men of science so readily fall, of obscuring my meaning by unnecessary technicalities."

1862. These were the lectures that the English novelist, George Gissing, would later satirize destructively in a set of novels ranging from *Demos* (1886) to *Born in Exile* (1892). Gissing had stalked Huxley intently, not physically as a person, but simply in the London Library, during his lifelong search for novelistic material that was unambiguously modern. Public lecturing designed to bridge the class divide was exactly his kind of subject. In *Demos* Gissing created a situation in which the idealistic lecturer to working men was tested by the manner in which his private life happened to develop. Did he want to devote his life to philanthropic working class lecturing or did he want to marry a middleclass woman and set up home appropriately in the conventional manner expected by her family? Of this earnest, but shallow popular lecturer the narrator's acid comment was that "the bent of his mind would have led him to natural science", since "he knew nothing of literature, poetry or history".[20] Though Huxley was still alive when these novels were published, innumerable references show that he was Gissing's model for characters historically trapped between philanthropic good intentions and the materialistic demands and expectations of British middle-class life. When Huxley chose to devote much of his life to the task of interpreting science to uneducated working men, he was creating a situation for himself similar to that described by George Gissing. This is not to say, of course, that Gissing was right and Huxley wrong, since the educational movement aimed at the "intelligent artisan" became even stronger in the twentieth century, not least in the Workers' Educational Association where Huxley's work was frequently studied and celebrated.

Huxley's preferred strategy for going about his business as popularizer is worth noting. In the "Advertisement to the Reader" he indicates clearly that the book is a popular introduction to the subject of man's place in nature, where the procedure can be legitimately different from the procedure needed for scientific

---

[20] *Demos a Story of English Socialism,* in three volumes, (London: Smith, Elder & Co., 1886)

articles written during the same period. During the years 1860-1863 Huxley published at least five articles on the same subject that are not referred to at all in "Evidences", which, however, does not mean, or does not only mean, that in this book there has been dilution for the wider audience. It means that he is approaching the matter in an entirely different way. Although in the book as published he allows himself discursive liberties that would be inappropriate in peer-reviewed articles, these liberties are useful to us in exposing the problems he had yet to solve. He could only work with what he knew at the time, which he, like many of his colleagues, did not think was a reason for not putting pen to paper. Sewn into most copies is a frontispiece depicting the skeletons of the gibbon, the orang utan, the chimpanzee, the gorilla and man with the caption "photographically reduced from diagrams of the natural size (except for the gibbon, which was twice as large as nature), drawn by Mr Waterhouse Hawkins from specimens in the Royal College of Surgeons".[21] Huxley had not seen a live specimen of any of the animals mentioned except the human being. In the early part of his career he was dependent on the facilities of the Royal College of Surgeons (of London), which meant that he could see the skeletons that interested him so much, but rarely, if ever, the live animals that the skeletons represented. This is by no means to fault him since other institutions, familiar to us now, were either in their infancy or did not yet exist. This is why R.M. Ballantyne in the boys' adventure book called *The Gorilla Hunters,* also published in 1863, can enjoy giving the measurements of the gorillas his heroes encountered in West Africa. His information was more up to date than Huxley's, which is perhaps why some have asked the question: was Huxley just a closet scientist without any inclination to work in the field? Did he desire to see specimens of the animals he was writing about? As a scientist, what did he consider to be evidence? Well, the

[21] Royal College of Surgeons, which was Huxley's academic base in London for many years.

answer to this is fairly clear. If he had any thought of gathering his own data in the field, the desire was well disguised.

Whereas Ballantyne in the single scientific chapter that was a feature of many of his books simply refers to his main source of information on gorillas, Huxley indulges in a long introduction listing every mention of apes and ape-like creatures he has come across in other people's books, a list which reflects many hours of work in the library of the Royal College of Surgeons and in his own study at home. The importance to Huxley of his home life, its essential character and qualities, and the long, solitary hours he spent in his study have perhaps not been fully understood or recognized. The long, text-based introduction of fifty six pages, to the extent the reader can be induced to bother with it at all, is actually a subterfuge designed to assure the reader that the purposes and procedures of *Evidence of Man's Place in Nature* will be scholarly but not didactic or dangerous, a subterfuge similar in method to the woods of Dunsidane being used by Malcolm to disguise the size of his army as he approaches Macbeth's castle. The castle Huxley was attacking, with the intention of complete demolition, was occupied by wrong-headed clergymen and unemployable, possibly titled, landed aristocrats who, if educated at all, were probably inane graduates of Oxford or Cambridge more given to playing cricket, he once said scornfully, than reading books. No antidote has yet been discovered for the medical condition whose symptom is hatred of Cambridge and Oxford Universities, prevalent among more than a few people who have never been to either place. Huxley never played cricket, not indeed any other sport, and did not subscribe to the old adage "mens sana in corpore sano". Though he passed Lord's cricket ground on his way home to St. Johns Wood, there is no record of his ever going in. He never came to understand the benefits of physical exercise. He thought the human body was a machine. If it broke down, someone else would repair it.

It looks as though Huxley's geologist friend, the distinguished and ever urbane Charles Lyell, when he read the proof sheets of the book, and perhaps also the manuscript at an earlier point in time, had urged caution, what Huxley had at first written being too blatantly, too aggressively anti-religious, he thought, for the piece to have the effect Huxley wanted. In August 1862 he wrote: "I am persuaded that the best way to carry people with you is to write as though you were not running counter to their old ideas. I know that it is easier to preach than practice this".[22] A little later, perhaps in October, he wrote: "I am rather uneasy about your *Evidences* after what you said yesterday. I know the book might have a great sale & will I hope in spite of anything you may say about the unknowable, but if the work is in sheets not quite thrown off do let me see it that I may try & suggest something of the *suavitur in modo"*.[23] To this same letter he wrote a postscript: "I hope you send none of these dangerous sheets to press without Mrs Huxley's imprimatur". Charles Lyell was not objecting to Huxley's main thesis, which was that human beings had not been created separately by God as described in the Christian Bible, but had simply evolved from the ape, albeit in a process he was as yet unable to describe. Lyell was simply objecting to Huxley's belligerent strategies which he considered counter-productive.

In the fortress that Huxley was attacking were enemies who believed human beings were more allied to angels than apes, that they had immortal souls and that they were the special, independent creation of a Christian god, and therefore not part of evolutionary nature. Notwithstanding the rhetorical strategy, however, whether modified in consultation with Lyell or not, Huxley's intention has already been stated in his title. Man is not different from other forms of life. He is not one of God's creatures akin to saints and angels. He will never sit on the right hand of God in heaven because there is no such place as heaven. On the contrary, he is part of "nature" and is observably a member of the same class as

---

[22] ALS: Lyell to Huxley. 9 August 1862, Imperial College. This is the letter in which Lyell says of their colleague, Owen, "he must be nailed".

[23] ALS: Lyell to Huxley, [October 1862?]. Imperial College

the anthropoid apes and their immediate relatives. The main part of the book (pp. 57-118) is devoted to demonstrating, on anatomical grounds, that this is the case.

For the purposes of this Introduction, where the main thesis is that, when he went to America, Huxley took with him both academic and psychological problems unresolved in his earlier work, it is important to be clear about where he stood, that is, what he understood and did not understand either in 1863 or during the period up to 1876 when he sailed across the Atlantic. His position is stated pretty clearly on pp.105-109 where he identifies unambiguously what in Darwin's work he can agree with and what not agree with. "Our acceptance of the Darwinian hypothesis must be provisional", he says, "so long as one link in the chain is wanting".[24] In the next paragraph he says: "I have put this conclusion as strongly as possible before the reader, because the last position in which I wish to find myself is that of an advocate for Mr. Darwin's, or any other views". He likes Darwin's approach because "it is not inconsistent with any known biological fact", and because, he surely no longer needs to say, it is unmuddied by religion, but cannot go the whole distance with him because he cannot yet see how any anatomical structure could be modified, or could ever have been modified. This is the problem he will carry across the Atlantic.

In this preliminary account of how Huxley thought, the word "link", as used above, is important, because right here in *Evidences as to Man's Place in Nature* is the source of the concept of the missing link. Although Huxley understood the idea of anatomical or structural modification over time, he still, at this point in his career, could do nothing other than look for sequences of physical change in which present "types" indubitably developed step by step from past types. He approached the question as a comparative anatomist not as a biologist or zoologist. Meanwhile, the idea of the missing link passed into popular culture and mothers scolded naughty children with imprecations like "Don't ever do that again. You're a terrible little monster, you missing link you!" There were a few

---

[24] *Evidence as to Man's Place in Nature (*see pp. 105-109)

decades when the idea caught the public imagination without there being any clear understanding of what it implied. Meanwhile, on a more serious level, the outright belligerence about which Charles Lyell had cautioned Huxley had attracted the attention of others. Taking note of Huxley's "Lectures on the Elements of Comparative Anatomy", as published in the *North America Review*, William James took the occasion to refer to two other works that he called "Six Lectures to Working Men on the Origin of Species" and "Three Essays on Man's Place in Nature", taking exception to Huxley's "rude & even malignant tone" towards Professor Owen. "It is a state of things discreditable to Science", James said, "when, to use the words of an English critic, people go to Professor Huxley's lecture room in somewhat of the same spirit as those with whom they would flock to a prize fight".[25] Looking back over the years, one can see that, psychologically, Huxley always needed enemies. How could he be right without someone else being wrong? He faltered when there was not an enemy in his sights, as happened at least once in North America.

In what sense then, at this stage in his career (the late eighteen sixties) was Huxley a true scientist, or, if a scientist, what kind of scientist was he? How would he have looked, it may not be impertinent to ask, to the research officers of a major American university, say, the Vice-President (research) or a Faculty Promotions and Tenure Committee? In one respect, his curriculum vitae would have appeared impeccable since he had published more than a hundred articles, a high proportion of them in refereed journals such as the *Philosophical Transactions*, the *Proceedings of the Linnean Society*, the *London Medical Gazette* and so on. Though a caustic member of any review committee might observe that the majority of these articles were more descriptive than analytic, with the implication that not much had been added to the sum of human understanding, the output was undoubtedly impressive, so that when employed he would be secure in his job. Huxley was not, however, too obviously employable,

---

[25] Houghton Library, Harvard

partly because the openings were few and tended to be specialised, and partly because he was determined to remain in London where the competition was intense. When he was appointed as lecturer in palaeontology in 1854 at the School of Mines, his misfortune was that Sir Roderick Murchison was appointed the Director the following year. Murchison, being a plain sort of fellow much addicted to common sense, thought that a lecturer in palaeontology should know something about palaeontology, so, seeing that Huxley did not, assigned to the new member of staff various tasks with which he was at first ill-equipped to cope. The resulting tension between the two men—the one aged 62 in 1854 with a substantial, at that time unchallengeable research record, the other aged 29 with a better than average research record but with his main work still in front of him—was lifelong, Huxley regarding Murchison as pompous, ineffectual and absurdly of the old school, Murchison regarding Huxley as a self-centred upstart whose knowledge was not yet adequate for the job he was supposed to be doing. As was, and is, so often the case in Britain, the consciousness of class also acted as a divide. When Huxley went to Murchison's house to have dinner with the King of Sweden, he subsequently told Hooker how distasteful the occasion had been for him. He preferred his own wife's "tall teas", as he called them, perhaps followed by a dram of Glenlivet, to any formal dinner party with footmen and old-world ceremony. Did this make him the misfit a promotion and tenure committee might see as a potential trouble-maker whose promotion should be delayed, or was he a man of exceptional talent still finding his way in the world? In the present book the latter interpretation is being preferred.

When all is said and done, however, the nub of the matter is that the American publisher, Appleton, wanted Huxley on the firm's list, not because he was anti-religious, nor because he was known to be a friend and ally of Darwin's, nor because he was such a prominent wheeler-dealer in the London institutional world, but simply because of his known strong belief in science as a way of life. It mattered not to Edward Youmans, acting as Appleton's agent, whether Huxley was a good or unexceptional scientist, though he saw him as well-established and well-connected. It was the belief that mattered. Youmans could readily relate to Huxley's convictions about the importance of science in modern life because they were convictions he shared. Appleton would eventually publish twelve of Huxley's books. At first Huxley thought that a relationship with Appleton would be some sort of peripheral or secondary part of his professional life but he was swiftly corrected. In the present book there is therefore a brief chapter on the inaugural years of Appleton's publishing venture called the International Scientific Series, because it was during the establishment of the Series that Huxley's American interests began to be consolidated. Once that has been established, the central and longest chapter of the book can be devoted to Huxley's visit to the United States, a visit packed with highly significant activity, all of it having to do with the social and intellectual importance of contemporary science. The final chapter constitutes a discussion of how things stood with Huxley, both as a scientist and as a human being, at what is being seen as one of the turning points of his career.

CHAPTER ONE

Huxley and the International Scientific Series

This brief, first chapter will be devoted to the founding years of the International Scientific Series because it was at this time (1870-76) Huxley began to work closely with a number of new American colleagues, all connected in one way or another with the publishing firm of Daniel Appleton & Co., by then based in New York. Although some of his work had already been published in America, he was not yet sure how best to relate to American professional visitors to Britain, partly because he could hardly help but share the then current patronising, imperialistic British attitudes to other peoples wherever on the globe they happened to be, partly because he was literally ignorant of how things were done in North America, and partly because his view of North American science was as yet quite blurred and indistinct. Despite the administrative and personal ups and downs that will shortly be described, Huxley's labours on behalf of this bold, imaginative publishing venture to some extent prepared him for the actual visit to America he would make in 1876, and indeed may to some extent have provided the reason for his making the trip. At least, when Huxley and his wife went to America, there were people there who already knew them. Furthermore, because Huxley was located for better or worse in one of the major eras of the printed book, *after* the period when hand-printed books had dominated the market but *before* scientists switched, generally, from printed books to highly specialized articles in professional journals, it was perfectly sensible from his point of view that any chance to cultivate relationships with major publishers should be seized upon. However, there are a few matters that must be mentioned first, rather in the way that a new building lot has to be cleared before construction can begin.

In the English village where this book about T.H. Huxley is being written, there was, in 1871, when the International Scientific Series was launched, a population of less than three hundred men, women and children, with five shoe makers, two tailors, an inn that brewed its own beer, a church recently enlarged in accord with an Act of Parliament that required there to be pew space for the entire population, and a charity school enjoined to educate twelve local boys and twelve local girls. The school was exceptional. A high proportion of those who worked on the ten farms were illiterate. Though there was the canal by which heavy goods and coal could be brought to the village from the city and though once a week a carter drove a wagon and horses to the nearest market town fifteen miles away, for most people there was little practical reason or need for any type of communication with the outside world. The rural isolation and static way of life was in fact exactly as described in many of Thomas Hardy's novels. Huxley was a hugely ambitious, largely self-taught, upwardly mobile man of great talent. It was essential for him and those like him not to be trapped in such an environment, though he perhaps carried with him, when he moved into the city, the conviction that, because many of the men and women he had encountered in his early years were illiterate, ignorant, uneducated and often stupid, the only way to help them was at a very basic level.

Any person who wanted to make something of his or her life, whether educated or uneducated, had to leave a rural environment of this kind and the obvious move was to London, the centre of an increasingly dominant Empire requiring a high degree of literacy in order to function, since its functioning powerfully and efficiently depended on a constant, world-wide flow of communication, accurate despatches, unambiguous written instructions, and legal, commercial and administrative records kept scrupulously with pen and ink by a multitude of clerks on high stools. This was the literacy culture that generated long letters, long written reports, long sermons, long novels and long scientific texts, those who could read and write—the relatively small proportion of the population that was well educated—sharing a language whose grammatical and syntactical forms were, for them, well-established, fixed and strictly functional. It was still possible to talk of "correct" English because the dominant social class needed a reliable, and reliably shared, method of communication. It was to this Metropolitan culture that T.H. Huxley contributed the thirty eight books, written in the manner of the day, that are listed in Appendix One. Within a culture of high literacy, albeit in some respects a privileged and exclusive one, writing and publishing books was one of the principal ways by which an ambitious man could establish himself, though Huxley's interests extended way beyond privilege and social exclusiveness.

This first chapter is therefore about Huxley's lifelong preoccupation with printed books and his closely related enthusiasm for publications that could be put to the service of popular education and the social betterment of the uneducated. By the time he returned to Britain from Australia, the age of thirty, the widespread manufacture of introductory primers for the growing educational market was already in progress, as publishers' lists and advertisements clearly show. Of course, this by itself cannot be the reason, or cannot be the only reason, for Huxley's continuing to write books throughout his whole career. That there were other reasons will become clear as this narrative develops. For the publication of scientific results—the present remarks are confined to science—scientists had two options. They could choose to write articles for the growing number of specialist journals, often or mostly the publications of the ever increasing number of professional and academic societies—this was part but as life went on a diminishing part of Huxley's environment, or they could write books, in order, as they nearly always hoped, to reach a wider audience. Huxley did have this hope. Books were potentially more lucrative than articles.

Putting the matter in very general terms, the first modern period in which it was respectable for a scientist to write books was a brief one. Very few scientists would choose to publish their research results in book form after, say,

1900, while before that date there were certainly many who wrote innumerable articles and research reports without the possibility of writing a book ever being seriously considered. For Huxley to publish upwards of forty books indicates that he had made a career decision. For whatever reason, he felt the need to call out to the wider audience. He wanted to be heard. He chose to utilize what seemed a viable, socially acceptable medium for the dissemination, not of research results in the strict sense, but of ideas about the importance of scientific education for the health of the community in which he found himself living.

These remarks are of course not intended to detract from his articles, or from the work on which they were based. He wrote and published almost three hundred scientific articles and it was on these that his early reputation as a scientist was based, but the subject or theme for the moment is to establish the way in which his belief in the social importance of the printed text predisposed him to devote valuable time to the contemporary matter of book production—its technology, its marketing practices, the use that might be made of it in promoting a greater public awareness of the importance of science and the income it might generate, and in his case did. Significantly, in several of the paintings of him, he is depicted as carrying a book. When he gave three lectures in New York, he chose to do so without notes but with a book in his hand, which, however, he never opened. The book, as people say these days, was iconic or, as people used to say, symbolic of a cultural attitude that asserted the importance of being educated or, at least, literate. He had largely educated himself with the help of books. Others could do the same. Or maybe he thought they should.

From the very beginning of his career reading and writing lay at the centre of Huxley's exhausting daily routine, a routine which involved reading everything his professional contemporaries (his peer group) wrote and writing articles, reviews, reports and books on his own account. This was something he knew about. By 1871, he was undoubtedly in a position to use his publishing experience to help launch the International Scientific Series. This is why the present book about

Huxley's 1876 visit to North America starts with this preliminary chapter about his knowledge and experience of book production, as well as his closely related dealings with American publishers. Book production was at the very centre of his professional life. Indeed, it may be well to mention at the outset that it was very rare for him to give a lecture that was not later published. Consequently, in any study of Huxley, the production sequence from spoken word to publication is of the utmost importance, since it has both practical and psychological implications.

The American Initiative and Huxley's Role as Planning Consultant

The International Scientific Series was the brain-child of the American publisher, W.H. Appleton and his elderly but versatile professional adviser, E.L. Youmans, though neither of them at first regarded the scheme as international, since their initial focus was exclusively on the publishing of books in the English language in Britain and the United States. The primary commercial aim was to increase Appleton's market share in North America of books first written and published by scientists in Britain, the inducement being to offer British authors exactly the same kind of contract and the same royalties as those expected by American authors. The publisher's judgement, right or wrong, was that Europe was ahead of America in scientific research, so that there would therefore be a demand for foreign works. The scheme was designed to confront and override the practice of those American business rivals who published the texts of imported books without the permission of the original European author or publisher, this book piracy, as it came to be called, being perfectly legal or, at any rate, not illegal, in as far as no copyright agreement yet existed between Britain and the United States, and would not until the final decade of the nineteenth century. Between, say, 1860 and 1890, if a British author wished to be paid royalties when his book or books were made available to American readers, he had best choose Appleton as his publisher. If he did not, he could not expect an income of any kind from American sales. Huxley was involved in the Appleton plan from the very beginning.

The International Scientific Series, as originally conceived, depended for its implementation upon mid-nineteenth-century technology on both sides of the Atlantic. Travel across the Atlantic had become easier. The telegraph facilitated communication. And the printing of all manner of texts, from newspapers to scientific monographs, had been greatly advanced, technically, by refinements to the printing technique of stereotyping. A page of text still had to be prepared in the print-shop in the traditional or pre-electronic, old-fashioned manner, the compositor creating a page of type letter by letter, but once this had been done the letter-text information could be transformed into stereotype plates which could be used many times and in any part of the world. Because any number of stereotype plates could be created from a single setting of type, one set could easily be shipped across the Atlantic and used by Appleton, provided, of course, he had modern, compatible equipment and machinery, which he had. In other words, international collaboration between publishers had become easier. If a new book could be made to appear at exactly the same time in each country, the separate copyright laws of each nation would immediately apply, and both authors and publisher would be protected from the piracy that was then rampant.

For the sake of the narrative that follows, it is worth noticing that, even before the International Scientific Series had been proposed and launched, Huxley had already benefited from this particular technical advance and the advantages carried with it. He had been in touch with the firm of Appleton from 1863 (at least), the year in which *Evidence as to Man's Place in Nature* appeared in both countries. A few years later, the *Manual of the Anatomy of Vertebrated Animals*, published by J & A Churchill in London in 1871, was stereotyped for Appleton, the American version, that is, the first edition in the United States, carrying the same date of publication. A close examination of the American and British editions of the manual side by side shows them to be textually and typographically identical. In other words, Huxley by 1870 or so was already familiar both with the technology of stereotype printing for the sake of simultaneous publication on both sides of the

Atlantic and with the financial benefits that were likely to result, and in his case did. He had also had dealings with Edward Youmans, Appleton's agent, having contributed both to Youmans' *The Culture Demanded by Modern Life* and to his *Elements of Physiology and Hygiene* in 1867 (see Appendix One).

In 1871 both Youmans and Appleton travelled to Europe to promote their plans for international collaboration, Youmans travelling first to prepare the ground, and Appleton following later in the summer. Dr. E.L Youmans (1821-1887) was a deeply committed scientific author, well known in American academic and publishing circles, who had previously been the editor of *Appleton's Journal* and who fervently believed in the educational importance of modern science, in exactly the same way as Huxley did. He was relying on this compatibility. In his collection of papers called *A Century of Science* (1880) John Fiske recalled how his friend had been completely blind for thirty years, but then in 1851 recovered his sight in one eye, this event giving him a new lease of professional life. Youmans had then written books intended for wide audience such as *Class-book of Chemistry* (1852) and *Handbook of Household Science* (1857). Since resigning his editorship, his main employment for D. Appleton & Co. had been reading foreign scientific works, so by 1871 he was well acquainted with what was being published by scientists in Europe, and certainly in Britain. He had already met John Tyndall during Tyndall's visit to the States a year earlier and he had corresponded with the other British scientists already being published by Appleton including both Darwin, whose *The Descent of Man* was beginning to sell well in the States, and also Herbert Spencer, whom he had first met while on his honeymoon in 1861 and with whom he had almost certainly discussed his ideas about science publishing before sailing for Europe in 1871.

As the well-informed, well-regarded representative of a powerful publisher Youmans was able to campaign energetically for sets of ideas he knew to be acceptable to those advanced thinkers such as Huxley, who saw the need for an educational revolution that would keep pace with technological, industrial and scientific developments throughout the western world. It would truly be a revolution if educationists could be persuaded that science was more important than the Greek and Latin classics that still dominated the curriculum in many schools and universities. Reasonably enough, Youmans saw Huxley as an important ally. In Volume Two of his *Essays Historical and Literary* (1902) John Fiske recollected the happy days during the early eighteen seventies when Youmans first met Huxley, Spencer and Tyndall in London. He had gone for dinner in Spencer's bachelor quarters. "Spencer's omniscience extended to the kitchen, and as composer of a menu neither Careme nor Francatelli could have surpassed him" (p. 204). He also attended Huxley's "tall teas", enjoying both the boisterous, friendly company—"of mirth and badinage there was always plenty"—together with the occasional glass of "Rhenish" or dram of Glenlivet, and also the long smoking sessions in Huxley's small study, which was where he one evening read the synopsis for a book to be called "The Dog", a book which was never written. He had likewise been warmly welcomed in John Tyndall's room on the top floor of the Royal Institution on Albemarle Street in London. These were the heartening, encouraging experiences that Youmans had, to the extent that Fiske said Youmans had come back from England bubbling and brimming over with racy anecdotes about philosophers and men of science" (201). It was because Youmans had such pleasant experiences and good talks with Huxley, Tyndall and Spencer during this first business visit that he subsequently sought their advice as the plans for an international publishing venture began to take shape. During that summer it was gratifying that scientists and publishers on both sides of the Atlantic had been able to get on with each other in such a comfortable manner. The launching of the Series was based on this fellow feeling as well as on the obvious elements of self-interest. Whether Series could be sustained or not no-one in 1871 could yet tell.

Just before Appleton arrived in Britain, Youmans tried to have him elected to the Athenaeum, which club he knew was the haunt of eminent British scientists.

Could Huxley get him in? What he says about his employer in his letter to Huxley is of some interest.

> Mr. W.H. Appleton is a man of quiet but powerful influence in this country, as is illustrated by the fact which I state to you *confidentially* that he has been consulted by our government regularly whenever the question of International copyright has been brought up. He is not an opponent to the policy; and is ready to allow English authors their rights whenever that can be done.[26]

Later in the letter came the persuasive sentence: "if it can be managed it will carry with it important material advantages". Huxley was interested.

Failing at the Athenaeum, Huxley got Appleton and his son elected to the Savile Club. At the same time, they found appropriate, conveniently located accommodation in the St. James's Hotel, using the club and the hotel as their headquarters. With their base camp established, as it were, they began to look for working relationships, not just with Fellows of the Royal Society of London, but also with that caucus of intellectuals lending distinction to University College London at that time. Having taken advice on how science was regarded in Britain, they aligned themselves with those whose view of science, and knowledge generally, was modern, utilitarian and secular. Among the many meetings during the early summer of 1871 was one at which he met Charles Darwin. With Youmans he went for a "juicy and jolly" lunch at Erasmus Darwin's where he met Charles, Charles' wife and two of their children.

Youmans later reported that it was Charles Darwin who "insisted" that the proposal for a new Anglo-American publishing programme, if it involved scientists, should be canvassed at the next meeting of the British Association which was to be held in Edinburgh in August.

---

[26] ALS: Youmans to Huxley, 28 April 1871, Huxley Papers at Imperial College, London: Vol. 29, pp. 259 ff.

Youmans took Darwin's advice and convened a meeting in Edinburgh to discuss the launching of the Anglo-American series. Obtaining a room in the ancient quadrangle of the Old College as it was then, he met a group of scientists which he told Appleton included Huxley, Carpenter, Balfour Stewart, Archibald Bain, Herbert Spencer and Dr. Lindsay from Perth, with whom Youmans was already in negotiation. Huxley explained to the group that he had done well financially from being published on both sides of the Atlantic, though he still used the words "reprints" for the American publications, which is probably one of the several factors that have allowed some people to believe the original impetus was British. John Tyndall was not at the meeting so presumably became a member of the advisory committee on Huxley's recommendation. Huxley and Spencer were still getting on well with each other. Their tragic, terminal, probably unnecessary dispute lay a few years in the future

It was obviously vital that Huxley, Tyndall and Spencer should support the Appleton/Youmans proposal once they had been consulted, not just because of their influence in the scientific community as a whole, but also because they could talk at first hand about the advantages of dealing with Appleton. Who better could lend the scheme such strong support? Youmans got on particularly well with Spencer with whom he would play billiards every night after dinner when they returned to London. Not only did Spencer at that point in time have an even greater experience of international publishing than Huxley and Tyndall, he also shared Youmans' convictions about the social sciences which is where they both thought science had its greatest public impact. From this point of view—the vantage point of the avowed modernist in 1871—sociology, psychology, jurisprudence and politics *were* sciences.

From the time of the Edinburgh meeting, an important defining aspect of the proposed Series was organizational. The visiting Americans absolutely had to have the feeling, when they returned to the States, that they had won the support of the British scientific community and that they could rely on the people they had met, especially Huxley, Tyndall and Spencer, to give them reliable advice on who should be published and who not. Had they believed there were people in the States who could give them equally sound advice, they would have sought for it at home, which of course in many cases they perhaps did. But even if they were doubling their bets, as it were, they could reasonably expect the scheme to work more smoothly if British scientists were involved. It was a sound diplomatic move to consult established scientists before the scheme was launched, though they would need to find a British publisher who was willing to take the commercial risks any scheme of this kind would involve. All the books in the International Scientific Series, as it was viewed at that time, were to be manufactured in Britain, as already explained. As a consequence of these first negotiations, the International Scientific Series was soon being spoken, in some quarters at least, as a publishing venture run for the authors by the authors, which is what Huxley, for one, thought it was or was determined it should be.

Although the time-span was short—just a few months between the forming of the advisory committee and the signing of an agreement with a publisher—there was during the interval a great deal of enthusiastic, idealistic discussion among scientists in Edinburgh and London, discussion which lacked input from anyone who knew about the manufacture and marketing of books. This may have been because the first London publishers to be approached declined the opportunity to be involved. Although the interval of time was a brief one, later events showed it to have been significant. Because the three British authors who had formed the ad hoc committee were already being published by Appleton, who was anxious to keep them on his list, they very understandably came to think that their advisory role was in effect a decision making role, the more so because of their involvement in the drawing up of the initial contracts, which were to have special features never adopted by other publishers. It is fairly certain that Huxley was the prime mover in this development.

What quickly emerged as one of the most important items on the English committee's agenda was the requirement that an author who could find time to write a book should be able to anticipate a reasonable, immediate income from it. Understandably, the British scientific authors involved wished to capitalize on the American initiative to the fullest extent possible, perceiving clearly enough, they believed, that if production costs were reduced by the British and American publishers having the advantage of stereotype printing, the profit margin should be greater. Huxley thought that some or all of the increased profit should go to the authors. No doubt after much talk, it was agreed in principle that in Britain an author should be paid for each new "edition" of a book at the time of publication, that is, before copies had been sold, where an "edition" was defined, not in the conventional manner, but as any set of 1250 copies, whether printed at the same time or not, while in the United States that same author would be paid a 10% royalty on the actual sales of the book as manufactured in the States. This unusual provision resulted in difficulties that will be described and discussed later. They were difficulties not anticipated in 1871, at least not by the people Youmans is known to have consulted. If a publisher could be found willing to pay authors the equivalent of a 20% initial royalty, he at least would understand, even if the authors did not, that he was postponing the time when he could expect substantial or even adequate profits, so the theory was that long-term profitability would have to be guaranteed by the eminence of the authors. Youmans had been warned, before crossing the Atlantic, that he could not count on his plan being accepted by any established London publisher, and now the recommendations or requirements of the committee made the task of finding one even more difficult. He was lucky to make contact with someone willing and able to accept the unusual risks.

Henry Samuel King had been a bookseller, then a partner of George Smith of Smith, Elder and Co. (the two men married sisters) with a quarter share in the business. When the partnership was dissolved, he agreed not to engage in publishing for three years. Nevertheless he had prospered in business before returning to publishing in 1868. By 1871, he had succeeded in extending his business activities, trading as Henry S. King & Co., "agents and bankers", specializing in transport to and from India, with branch offices in Liverpool, Southampton, Bombay and Calcutta. Associated with the main part of his business were several publications. Among them were *The Overland Mail*, "A Summary of Home Intelligence for India, China and the East", *The Homeward Mail and Official Gazette*, "A Summary of Intelligence from India, China and the East", *The Week's News*, "A Family Newspaper for England, the Colonies and the Continent" and *The Day of Rest*, "An Illustrated Weekly Paper for the Sunday Reader". His principal office was at 45 Pall Mall, his shipping and supply office was at 65 Cornhill, and his baggage office was in Finsbury. By the time he returned to publishing he commanded an extensive business network with much experience of the transport of people and goods throughout the world. Perhaps partly because of this, he chose not to be a specialist publisher but to build up a general list designed to be of interest to the communities he served, particularly in India. King had made his own way in the extremely lively and competitive British publishing world, was accustomed to making his own decisions, had the wide experience of a well-established business man, and would therefore enter any new enterprise with his eyes open. He certainly had the strength of character to match anyone who attempted dominate or bully him. Though the idealists who planned the Series did not yet know it, the stage was set for a classic encounter between two men of equal determination and professional conviction.

Appleton, Youmans, Huxley, Tyndall, Spencer and King met at the St. James's Hotel in London on 14 August 1871 for a final discussion of their provisional plans before Appleton returned to New York. Whether in private discussions or in consultation with others, Appleton and King had arrived at a clear, hard-headed business understanding of how the scheme might be implemented in practice. A memorandum of agreement had been drawn up which mostly recapitulated what had been agreed during the summer. Huxley, Tyndall and

Spencer would constitute a "consulting committee". A series of "scientific monographs" would be produced and would be called "The Anglo-American Series". Henry S. King would be the English publisher, having agreed to publish the series on terms which Huxley, after negotiation, considered favourable to the authors, specifically that, while the publisher would bear the whole cost of production, the author would receive a "royalty" on or before the date of publication equal to one fifth of the selling price of the first "edition", and that the same provision would apply to subsequent editions where, however, there would be a six month delay before payment was made.

These terms need to be explained. First, the use of the word "edition" here and elsewhere in the planning documents was unusual, perhaps even eccentric. What Huxley had in mind, as at this point the most active member of the committee, was not a set of books, whatever the number of copies, deriving from a single setting of type (that would be the normal definition of an edition), but simply the printing of 1250 copies, the second set of 1250 copies being called the second edition, and so on, the figure 1250 always (in these early days of the English series) being the controlling element for the purposes of payment. Huxley was trying to improve on the normal or traditional situation in which an author, unless famous, has usually to wait until books have been sold before an income can be expected. Why write a book except for money? That sometimes seems to have been Huxley's entirely sensible attitude. For him a book was simply a commodity for the market place, not the medium for the profoundest types of reflection or self-expression. Second, according to a well-established publishing convention, 1250 copies would count as 1000 for accounting purposes, the publisher having the disposal of the remaining 250 copies from which the author could expect nothing, since he had already been paid £50 which was calculated as one fifth of the retail price of a thousand copies. Even though part of the 250 copies would be distributed for review and presentation purposes, he would have a little more leeway, financially, than perhaps the committee realized. A gap was beginning to open up between the publisher, who knew a lot about publishing, and Huxley, as the authors' self-appointed representative, who knew very little about publishing, beyond the fact that all authors like to receive payment for their labour—preferably substantial payment. For his part, Appleton agreed to purchase copies of the stereotype plates, together with electros of any woodcuts (for the illustrations) directly from King, undertaking to give "all English authors dealing with them direct in this manner, the same rights as American authors".[27]

The agreement of 1871 had put Youmans in the position of being able to prepare a printed notice about the Series for distribution at the Edinburgh meeting of the British Association. He had a completely clear idea both of the purpose of the Series and how it was to be managed. The initial planning stage had been completed to his satisfaction in a manner that would strengthen the educational role of science, while bolstering Appleton's position in the market place. He knew that Huxley was with him on both counts. For the whole of his career Huxley remained convinced that science, not the humanities, was the essential ingredient in a good, modern education, that students should be taught the rudiments of science at the earliest possible age, and that the teaching should be through books as well as by demonstration.

As for the market place, he showed no resistance to the thought of becoming as wealthy as possible. Being published in America might be a great help in this regard. Youmans' notice in part read as follows:

> The general aim of the Series will be to give authentic, popular expression to the latest advances of thought on the leading subjects of progressive enquiry. The recent and more progressive steps of physical investigation will come within its scope, and those interpretations of nature which have undergone a marked revision within a recent period. Yet it is desired to give especial prominence to those various branches of biological,

---

[27] There is a copy of the agreement dated 14 May 1872 at Imperial College.

> psychological and social science, which help to a better understanding of the phenomena of human nature and the economy of human life.[28]

He believed, as Huxley did, that popular, influential and educationally useful books could be written that would be "thoroughly explanatory and expository in character", that they should be free of "technicalities" that only specialists could understand and that, if they carried "the weight of authoritative names" they would sell well, not least in the United States "where the interest in science, especially in its progressive aspects, is also becoming earnest and general".

> It is designed to consist of compendious scientific treatises, representing the latest advances of thought upon subjects of general interest, theoretical and practical, to all classes of readers. The familiar phenomena surrounding Nature, in their physical and chemical aspects, the knowledge of which has recently undergone marked extension or revision, will be considered in their latest interpretations. Biology, or the general science of life, which has lately come into prominence, will be explained in its leading and most important principles. The subject of mind, which, under the inductive method and on the basis of its physical accompaniments and conditions, is giving rise to a new psychology, will be treated with the fullness to which it is entitled. The laws of man's social development, or the natural history of society, which are now being studied by the scientific method, will also receive a due share of attention. While the books of the Series are to deal with a wide diversity of topics, it will be a leading object of the enterprise to present the bearings of enquiry upon the higher questions of the time, and to throw the latest light of science upon the phenomena of human nature and the economy of human life.[29]

This statement was a reflection of Youmans' own convictions and at the same time was probably not unfaithful to what had been agreed in Britain.

Huxley never showed very much interest in the social sciences, indeed was not slow to object when the word "science" was used in ways of which he did not approve. In 1871 and 1872, however, he was at first complacent about

---

[28] Huxley Papers, Vol. 30, p 98, 28 December 1871, addressed to "British Scientific Authors" and marked "private and confidential"

[29] ibid.

Youmans' extension of what the committee had agreed. For Youmans the social sciences, if "studied by the scientific method", were from the inception of the Series seen as an important and equal part of the enterprise. Herbert Spencer agreed with him. As the social scientist among the originators of the Series, Spencer was particularly influential during these early years when the theory of the enterprise was under discussion but before very many books had appeared, but Youmans had also taken advice from American colleagues like John Fiske at Harvard, whose support Youmans found useful. Whatever initial reservations some members of the British academic community may have had about the status of sociology, psychology and political economy as respectable academic disciplines (and the prejudice against them remained strong in some English universities until well into the twentieth century), these reservations were not shared by Youmans, Spencer and Appleton. Of particular importance here is Youmans' phrase "the latest advances of thought upon subjects of general interest". The key word here is "advances". Intended was any advance involving new knowledge which consolidated the supposedly modern freedom from illogicality, speculation, metaphysics and superstition, and from the power of the established churches, where opinions were based on evidence or "fact", not upon revelation or inherited wisdom. This lexical transfer from "scientific" to "advanced" was problematic. In the long run any laboratory scientist either might not care whether his research results could be called advanced or not, or else might prefer not to be associated with interpretations non-scientists chose to read into his work. Youmans had his way, however, and was undoubtedly supported by Herbert Spencer.

A good example of what was meant by "advanced" among the enthusiasts who initiated the Series is provided by Sheldon Amos' *Science of Law*, published in London and New York in 1874 and therefore part of the first tranche of new publications Appleton hoped the scheme would generate. It is easy to see how acceptable to Youmans and Spencer this author must have been when they first discussed the publication of this book. Amos believed that, through the centuries, there had been a forward historical development and real progress both in Government and, in the broadest sense, the law, to the extent that the modern state had reached "a climax of intelligent legislation and conscious political life". Consequently it could be studied scientifically. Because society is "from first to last a subsisting organization made up of constituent groups reciprocally acting and reacting upon one another", the interpretation of its laws, once it has reached perfection, will be by "logical processes" which are "permanent and universal". Idealistically, he writes paragraphs like the following:

> Every influence which increases the economical dependence of States upon each other, every barrier which is removed from free commercial intercourse and free locomotion and communication, not only makes war more and more opposed to the obvious interests of all classes of persons in every State, but brings home more and more distinctly that even States apparently the weakest and smallest have a function to perform in the great society of States which, in dignity and importance, may vie with that performed by the most pretentious. Every measure that gives political enfranchisement and personal liberty to large classes of the population must tend to the disuse of the enormous European armaments which absorb so much of the wealth of all states, and which, by keeping up a fictitious and delusive inequality among States as well as nursing a childish passion for military glory, tends to a like extent to promote war. Again every move that unites together the mass of the population of different countries, by teaching them what is their common interest, by removing traditional prejudices, and by bringing them, through locomotion and mutual interchange of labour, face to face with each other, must tend to diminish the possibility of war.[30]

At first glance, passages of this kind are international but not strictly scientific, though dressed up to appear modern. Possibly the word "science" was imposed upon a text that already existed and was awaiting publication by some publisher still unidentified until Youmans made contact with the University College faculty.

---

[30] Sheldon Amos. See Leslie Howsam, "An Experiment with Science", *BJHS*, June 2000, Vol. 33, pp. 187-2007

For the moment, the book will serve as an example of what might be called the Youmans/Spencer view of "advanced" thought, a view at variance with the sort of science that was based on observation, measurement, experiment, further experiment, analysis and logic. The passage quoted above is transparently illogical and for this reason alone, though there were other reasons, unacceptable to Huxley. On the other hand, it has the attraction of being unambiguously Utopian in its conviction that nation states might prefer to adopt agreed conventions or laws of peaceful co-existence, and this despite the recent Franco-Prussian War, the incursions by force of the colonial powers around the globe, the sabre rattling of Russia in the Middle East and so on. However defined, the law was not then, and never has been a science, but the views of Sheldon Amos, an important member of the UCL set that helped shape the Series, were "advanced" in the sense promulgated by Youmans. We have here a hint of problems to come. Among those interested in the strong, truly international development of the Series, there were those who regarded science as an activity strictly confined within its own intellectual boundaries, its own methodology, its own languages and its own objectives, and therefore untranslatable and to a large extent unintelligible to the uneducated, and others who thought that, because science was by definition not a function of any political system or ideology, it was in its very nature separate and free from the circumstances that resulted in both domestic and international disputes. Science could bring people together because not contaminated by doctrine, believed Spencer and Youmans. Science was the willing handmaiden of the warlike, was the other possibility. During the early years, when the International Scientific Series was being established, the almost inevitable conflicts between scientists and social scientists were muted. The publishers were enthusiastic, so the Series developed rapidly. But Huxley soon had to declare himself as a person unsympathetic to the notion that sociology, psychology and politics were sciences. In other words there was a potential rift between Huxley and the other people immediately involved.

The first two volumes in the International Scientific Series appeared in 1872. Each was published simultaneously in London and New York as planned. The scheme worked. But, while the focus is on what Huxley did and did not do, or could do and could not, we now come to one of the crucial moments in the history of the International Scientific Series. In the latter part of 1871, after Appleton's return to the States, and after the British Association meetings in Edinburgh, what had started out as the "Anglo-American Series" was transformed into the even more ambitious "International Scientific Series". Youmans and Spencer crossed the English Channel in November 1871 to tell French scientists and publishers about the new enterprise. Whereas Huxley's interests were principally British, even chauvinistic, with a focus on his own personal participation in any scheme and the income it might bring him, Herbert Spencer was genuinely international in outlook, believed that a scheme that might work in Britain might also work in other countries, and had the energy to put the matter to the test in consultations with Continental publishers. By the time Youmans and Spencer went to France, however, Huxley had become seriously ill. Whatever the illness, his professional life was in disarray because, amongst many other problems, he now had not one but several unfinished books in his study. The pressure on him was intense. As a result of this breakdown, he was out of action, relatively speaking, for two or two and a half years from late in 1871, whether simply from a collapse due to overwork or from other associated medical conditions as well. It was during this period that Youmans and Spencer extended the scope of the Series to include France, Germany, Italy and Russia. In doing this they were acting on behalf of the publishers, not on behalf of the London committee whose mandate was simply to make recommendations on books offered for publication. Whether Huxley knew what was happening cannot be determined. Probably not. He was in any case too ill to participate.

Because Germer Baillière's publication of scientific books were so well known in Britain, it is highly likely that Spencer and Youmans went to him first.

He quickly agreed to participate in the scheme, a contract was drawn up to this effect and the terms of the Paris agreement were then used as the basis of negotiations in Germany, Italy and Russia. Because Baillière was well-known throughout Western Europe, the foundation had been established for the enlistment of other publishers.

His business in Paris settled, Youmans left Spencer in Paris and went on to Berlin, where he met a number of scientists including Helmholtz, Virchow and Du Bois-Reymond. These were senior, well-established scientists whose work Huxley would have known and who simply had to be consulted before any new scientific publishing programme could be established in their territory, or so Youmans had been advised. Hermann von Hemlholtz (1821-1894) was a physicist and physiologist, who had just moved to the University of Berlin as Professor of Physics. Rudolph Virchow (1821-1902) was a research physician and pathologist who in 1869 had founded the Society for Anthropology, Ethnology and Prehistory. Huxley would have known his books, including his handbook on pathology and therapy and he would have known about his opposition to Darwin as well. Youmans failed to establish any kind of rapport with the Prussian scientists he met at the University of Berlin. He described the Germans he met in Berlin as "cocky"—so sure of their superiority that they were not in the least interested in the translation of foreign works. That, anyway, was Youmans' interpretation of what occurred. Possibly the Berliners he consulted were simply not persuaded that the Series offered sufficient benefits to be worth the trouble, since established scientists had no difficulty in finding Prussian publishers who could negotiate foreign rights whenever necessary, and it is possible, too, that they thought being "popular" meant excessive simplification, as many other people would believe until the Series established itself. Berlin's resistance may have had an impact on the Series at a later date. For the time being, Youmans had simply not succeeded in this part of his mission. Having been uncomfortably cold in London, he was uncomfortably hot in Berlin, at least during the night. Depressed by the reception he had received, he went to bed early. "To bed! And such a bed—not a sheet, not a blanket, but only a kind of narrow feather-bed; they call it a *down* that you have to sleep under. And you *can't* sleep under it".

With Huxley's help, however, earlier advances had also been made to publishers in Leipzig through the good offices of Isidor Rosenthal at the University of Erlangen. Though Youmans did not speak German, he evidently managed to describe the Series in favourable terms. Because the groundwork had already been done, the Series could be described as a publishing scheme that already existed, indeed was already being implemented. European authors, he explained once again, would enjoy the benefits of publication in London and New York. Whatever arrangements they made with their own publishers at home, they would also be assured of copyright protection and foreign royalties, provided a work had first won the approval of the London advisory committee. Simultaneously, the publisher's own list might be strengthened by the inclusion of American, British and French titles. Back in London, more dead than alive, he said, Youmans was able to report to Appleton: "Brockhaus, the great house of Leipzig, will probably publish the Series, paying foreign authors seven and a half per cent". The contract between Brockhaus and King was almost identical to that with Baillière. It was a bold and logical step that Youmans and Spencer took after agreement had been reached in London, and the boldness was matched by conviction, since they at least, among the various people involved at this early stage, genuinely believed that whatever ideological differences might work to hold nations apart, science would over-ride them, being by definition international, non-political and rational. Books by French and German authors recommended for the Series would still have to be approved in England before they could be stereotyped for simultaneous publication in London and New York, and there would be the additional cost of translation. On the other hand, Baillière and Brockhaus would have the option of publishing scientific works that might not otherwise have come their way. In theory, despite a few logistic and administrative problems that remained to be overcome, it looked as though the scheme brokered by Youmans and Spencer would be beneficial to all concerned. Would Huxley approve of what they had negotiated? They did not yet know.

Just as an advisory committee had been established in London to help Henry King identify suitable subjects and authors, to assess the merits of both solicited and unsolicited manuscripts, and to advise him on whether foreign works were worth translating, so now during Youmans and Spencer's Continental tour, and subsequently in contractual negotiations between American, British, French, German and Italian publishers, it was proposed that each of the participating countries should adopt similar rules and procedures—the idea being that only scientists could ensure the type of quality control which a publisher might not exercise, it was thought, either from ignorance of science or because principally motivated by commercial considerations. At least one of the publishers concerned complied with this. In Leipzig the committee at first consisted of Professors Rosenthal, Czermat, Marguardin (sic) and Schmidt. It was anticipated that while a participating publisher might admit into his part of the Series, in scrupulously accurate translations, most of the books published elsewhere, he was by no means committed to admitting them all. Nor would an advisory committee be expected to recommend them all. Obviously it was important that a publisher should be protected from the obligation to publish books he did not want and desirable that he should have independent advice of his own if a national series were to function at a high level. Members of an advisory committee, when vetting submissions to publishers, would have the power to block both the publication of works by fellow scientists in their own country and also the translation of a foreign work if they did not like the original, irrespective of the way in which the book had been received in the country of origin. In 1871 and 1872 these were no more than theoretical considerations. Youmans, on behalf of Appleton, was trying to devise an international publishing arrangement that could be expected to work. Appleton for his part continued to see the advantages of being able to publish the works of foreign scientists in America. He trusted his agent to give reliable advice on what might most usefully be done. Because stereotype printing was fundamental to the plan's success, it at first seemed sensible that the quality control should be located in London. This gave the London advisory committee considerable power. Would the power be exercised judiciously and diplomatically, or in a chauvinistic manner that implied only the British were capable of sound decision making? Given such power, would Huxley have his say in a constructive helpful manner, or would he try to impose himself on the whole decision making process? Was he capable of working amicably and creatively with experienced, even distinguished colleagues? Those who had worked to set up the Series—Appleton, Youmans and Spencer—were soon to find out.

In New York, however, there was at first no reason for anxiety. General confidence was expressed in the firm wording of the advertisement that appeared in early American editions.

> D. Appleton & Co. have pleasure in announcing that they have made arrangements for publishing and have recently commenced the issue of a SERIES OF POPULAR MONOGRAPHS, or small works, under the above title, which will embody the results of recent enquiry in the most interesting departments of advancing science.
>
> The character and scope of this series will be best indicated by a reference to the names and subjects in the subjoined list, from which it will be seen that the co-operation of the most distinguished professors in England, Germany, France and the United States, has been secured, and negotiations are pending for contributions from other eminent scientific authors.
>
> The works will be issued in New York, London, Paris, Leipzig, Milan, and St. Petersburg.
>
> THE INTERNATIONAL SCIENTIFIC SERIES is entirely an American project, and was organised by Dr. E.L. Youmans, who spent the greater part of a year in Europe, arranging with authors and publishers.[31]

---

[31] Because this was a circular, many copies exist. The copy quoted was seen at Impreial College.

The Huxley Crisis

Though trained as a medical student, Huxley showed little capacity for taking account of his declining health or for understanding his actual physical condition, for which he often used the all-encompassing word "dyspepsia", because the symptom most evident on a daily basis was severe indigestion. His body simply stopped working properly. He felt out of sorts but did not understand why. Certainly he is not to be faulted for not understanding how his own body worked, in as far as the depression associated with nineteenth-century definitions of dyspepsia had not yet been successfully analysed. His contemporary, Henry Maudsley, also a contributor to the International Scientific Series, whose career had started at the Cheadle Royal Hospital, south of Manchester, was still prescribing beef tea and tennis for his psychiatric patients, while at the same hospital the staff depended on the wine cellar for the frequently-used treatment of conditions not yet treated by drugs (the drugs did not exist) and, generally speaking, not yet understood. Mrs Beeton identified dyspepsia as a failure of the digestive system, frequently due to "unhygienic habits in life, to faulty dieting and especially inadequate physical exercise". Then she says, tellingly: "but the state of mind also plays a frequent and important part, both in its causation and persistence", the state of mind being one of "worry" and "anxiety", or what nowadays might be called stress. Mrs Beeton's sound but sometimes difficult to follow advice, in various editions of her book *Household Management*, was that a person would be happier and work more efficiently if the whole digestive system could be induced to function properly. It often happens, though, that workaholics like Huxley either do not know this or prefer not to take sensible care of themselves because convinced they have more important matters to bother about. What bothered Huxley more than his own health was other people's insistence that he had a soul. He was sure he did not. There were still people who believed in angels and the creed of the Anglican Church. He would have to devote a lot of time and energy correcting them. So anxious had he become about the need to insist on the physical basis of human existence that he overlooked the need to look after his own body.

Not much is known about the medical advice Huxley received, so it can only be deduced from events, and then only sketchily. Possibly he would have been a happier man had he been able to take advice from a modern dietician, an idea which of course is mere fantasy. Dieticians in the modern sense were not part of the London scene in the eighteen seventies. As mentioned earlier, Huxley suffered a physical breakdown soon after the launching of the International Scientific Series, so his declining health may well have affected his judgement and his relationships with colleagues. Getting away from it all was seen as a possible palliative, a possible medical remedy for what he had to admit, and everyone else could see, was extreme overwork. He went to Egypt during a long leave of absence, though for a power-broker to be away from London was more likely to add to the stress than reduce it. He said he had not enjoyed being in Egypt. Dissatisfied with Egypt, he spent his next long holiday on the Continent, believing, like so many others before and since, that he could walk his way back into health. Then, a couple of years later, he crossed the Atlantic for the visit that is the main subject of this book. There is a progression here: Egypt, Europe, America. He had been failing to meet both his day to day and his long-term commitments. Something drastic had to be done.

It is fairly obvious that, whatever the medical or psychiatric diagnosis of his condition might have been, Huxley was certainly, also, the victim of what came to be called the work ethic. Getting established in London, had been, and still was, a struggle. He habitually, compulsively, endlessly, overworked. Like hundreds of other people in the often crucial middle part of their careers, he had nearly always felt bound to accept such work as came his way. Before 1863, in roughly the first ten years after his return from Australia, he had accepted writing and publishing assignments that had merely kept him going financially, while bringing him to the attention of his peer-group and other science writers with compatible attitudes. These are the first set of items listed in Appendix One. It is not for a moment being suggested, however, and this needs to be emphasised, that he was insincere in accepting such work, or that he did not do it well, only that he had yet to find his true metier or the direction he wanted his professional life to take.

During the period of ill-health, Huxley began to have difficulty with his own production schedule, the same problems as of course thousands of other writers have also experienced. There was so much he wanted to do. His professional ambition was limitless. There were simply not enough hours in the day. One sign of this is the preface to *The Anatomy of Vertebrated Animals* (1871) where he says: “A considerable portion of this book has been in type for some years”, which, charmingly, he gives as the reason for its not being up to date. There is a similar statement in the preface to *Critiques and Addresses* (1873) where he refers explicitly to his own breakdown.

If the years of extreme personal stress, career anxiety, general ill-health and breakdown are, roughly, late 1871 to perhaps early 1875, it is relevant to remember that in some of his later books he fell into the habit of seeking editorial and research assistance, as in *A Course of Practical Instruction in Elementary Biology*, where he acknowledges the help of his colleague, H.N. Martin. He more and more sought such help. In the preface to *A Manual of the Anatomy of Vertebrated Animals,* which is dated June 1877, he states that the work was started twenty two years earlier, but was interrupted by “a period of incapacity for any serious toil”. It is not being suggested here, or implied, that Huxley’s position as a writer of books was unusual or in any way culpable. Over the years many good books have been improved by research assistants and associates. It is easy to say a writer should write his own books but there are many ways of achieving this. Moralizing on the subject of another person’s difficulties would be unhelpful. The point being made is that Huxley’s participation in the day to day affairs of the International Scientific Series was interrupted by the progressive deterioration of his health at crucial periods of time between the launch of the Series and his visit to America. He was deeply involved during the planning period; he did not re-engage with the Series an a meaningful way until some years later, despite gallant attempts to do so, though it is worth remembering that even then the career problems still persisted to the extent that, continuing to feel under pressure, he more and more depended on earlier work and found it difficult to engage in new research. It will be seen that Huxley's book *The Crayfish,* which was number 28 in the London and New York Series, provides a vivid illustration of this vulnerability, if such it is.

The question of Huxley's health is important for this narrative, first because it affected the way in which the British part of the Series was developed, leaving it weaker than, say, the French part, and second, because it became one of the several reasons for his crossing the Atlantic.

Before these two questions are addressed, though, there are some general matters that need to be mentioned. To think about the development of the International Scientific Series once it had been set up means taking account of the time that elapses between the acceptance of a manuscript for publication and the book's eventual availability for sale. At first this time period was fairly short, partly because the organizers were so keen to see their plans in operation and partly because the authors of the first books to be published already had work on hand, which they might or might not be prepared to structure in the manner being required or suggested. Of the first dozen or so books published in London, at least two failed to meet the criteria Youmans had had in mind. These were John Tyndall's *The Forms of Water* (he wrote a book as though for children,—not very clever ones—showing that he had completely misunderstood what had been agreed) and Edward Smith's *Foods*, which was neither genuinely scientific nor up to date. Among the others, though, there were valuable contributions of the kind Appleton and Youmans had hoped for, including Walter Bagehot's *Physics and Politics*, Herbert Spencer's *The Study of Sociology* and Henry Maudsley's

Responsibility in Mental Disease. The first translation from the French was published in London and New York in 1874. This was E.J. Marey's *Animal Magnetism.* The first German work to be added to the Series was Oscar Schmidt's *The Doctrine of Descent and Darwinism* which was published simultaneously on both sides of the Atlantic in 1875. Then, just as the Series was building up a good head of steam, it shuddered to a halt. Books that had already been commissioned appeared in 1875 and 1876, including one from Italy, two from France and three from Germany, but there had been an administrative or quasi-political crisis in London for which Huxley had been largely responsible. Although the Series was beginning to prosper in France, Germany and Italy, and was therefore becoming truly international, only two British books were contributed to the Series between 1875 and 1879, and these were books for which contracts had existed before the crisis occurred.

Youmans' revised advertisement—the one mentioned above which was put out after the Edinburgh meetings—was accompanied by a list of scientists who had already agreed to contribute works to the Series. The full list reads as follows:

T.H. Huxley *The Races of Mankind*
John Lubbock *The Antiquity of Man*
W.B. Carpenter *The Principles of Mental Physiology*
Francis Anstie *The Influence of Narcotics*
Prof. Thistleton-Dyer *Later Aspects of Botanical Science*
H.C. Bastian *Cerebral Physiology*
T. Spencer Cobbold *Human Parasites*
Edward Smith *Food and Diets*
Alexander Bain *Relations of Mind and Body*
Alexander Geikie *The Framework and Sculpture of Mountains and Valleys*
Prof. Balfour Stewart *The Conservation of Energy*

W. Kingdom Clifford *Mathematics Interpreted to the Unmathematical*

J.N. Lockyer *Spectrum Analysis*

These were well-established scholarly and professional heavyweights. They were all distinguished so almost any publisher would want books from them. Within the list were five Fellows of the Royal Society, one Fellow of the Royal Society of Edinburgh, two MDs and several professors. Impossible to suppose they were listed without permission. On the contrary, discussions between authors and the publisher must have been in progress when the list was published. Books by three of the authors listed above, those by Smith, Bain and Stewart, were delivered promptly, presumably because the writing of them was well advanced before the crisis occurred, but the others either never appeared at all or were published much later than planned. [Lockyer 1878, Bastian 1880, Clifford posthumously 1885:]. Why did the other books on the list never appear?

The answer to this question lies in the functioning of the advisory committee and its relationship with the British publisher, or rather in the failure of the committee to function as originally planned and in the reluctance of the publisher to be directed, as opposed to being advised, either by the committee as a whole or by individual members in it. It quickly became evident to the publisher that members of the committee were not in agreement with each other when it came to the crucial business of recommending works for publication, as can be seen when their attitudes and actions are regarded individually.

It was at first Herbert Spencer who had the most frequent dealings with Henry King, to the extent that the publisher found it easy to assume, as Charles Kegan Paul later recollected, that Spencer spoke for the whole committee whenever he visited the publisher's offices in Cornhill. With Youmans, Spencer had engineered King's valuable contacts with Baillière, Brockhaus and Dumolard, and had also been instrumental in introducing King and Germer Baillière to the proprietors of the Russian journal and aspiring publishing house of *Znanie* with whom he had had earlier dealings. Spencer was at that time the most widely published member of the advisory committee. He made a point of keeping in touch with Youmans, who was indeed a friend of long standing, as Fiske's life of Youmans attests. There is nothing to suggest that he had anything but amicable relations with Henry King. He strongly believed in Anglo-American academic and commercial collaboration and he believed, also, in the educational and political value of the international entente he had almost single-handedly engineered, albeit with the backing of King and Appleton. Interestingly, when Spencer saw the way the wind was blowing, he took the occasion of Huxley's move into his new house to celebrate their long friendship with the gift of a drawing-room clock. "I have often wished to express in some way more emphatic than words my sense of the many kindnesses I have received at your hands during the twenty years of our friendship … there is no one whose judgement, on *all* subjects I so much respect, or whose friendship I so highly value".[32] These sentiments were about to be seriously tested.

Whereas Spencer played an active, positive role during the first years of the Series' existence, John Tyndall by contrast did next to nothing. He was not an active player in the scheme, unless in mostly undocumented dealings with Huxley. There is an undated letter of Tyndall's to Spencer which expresses his attitude to meetings pinpointing the quandary felt by many faculty members then and nowadays. Should a scientist be a good citizen and attend internal university meetings where the matters discussed were often more bureaucratic and mundane than intellectually important or, if possible, should he or she insist that remaining in the lab or the office doing one's own work was a better, more appropriate use of time? Tyndall thought that he had work in progress at the Royal Institution he did not wish to see interrupted. When involved in a set of experiments, he said. "I would not kiss the hand of the prettiest girl in London for fear of subsequent perturbation … you must count on me as an irregular; I cannot undertake any

[32] Spencer to Huxley, 1 January 1873, University of London Spencer Collection 791/15.

responsibility".[33] Whether in order to avoid the troublesome problems that were beginning to show themselves in the management of the Series, or in order to get on with his own work, as he said, he returned the £100 retaining fee and distanced himself from day by day workings of the Series.

In November 1873 Huxley had asked King about the advertisement of a book by W.B. Carpenter which he believed had not been approved or discussed by the committee. This was a technicality. He knew Carpenter had been expected to contribute to the Series, just as he had himself. Huxley had not yet written a book to the Series; now Carpenter had. Huxley, acting unilaterally, chose to confront King on a procedural matter and in one way or another prevailed, since no book by Carpenter ever appeared in the Series, though King did publish the work in question which, in the event, was perhaps too long and technical for the Series as originally conceived, though according to his own account Huxley could not have known this. Here we have the sort of incident that in academic circles tends to do the rounds in the common rooms and coffee shops when mere chatter and gossip can have a negative effect on the best of projects. Why offer a book to the Series, if your proposal might be treated in this way? An established scientist of the kind Appleton, Youmans and King were hoping to attract was unlikely to be delighted by the discovery that a person he had no reason to respect was sitting in judgement on what he had offered to the Series in good faith. Although Huxley claimed to be acting on behalf of the committee, there is no evidence that Spencer and Tyndall participated in this first confrontation between Huxley and King.

Only a couple of years later, the signs of strain and disagreement became more marked, as a letter from Huxley to Spencer makes clear.[34] Spencer wrote to Huxley about the way publishing decisions were being made. Huxley replied that he could not remember whether anything had been said to him about Carpenter

---

[33] ALS Tyndall to Huxley, University of London Spencer Collection 791/15

[34] Huxley to Spencer, ALS: Huxley to Spencer, 15 June 1875, Spencer Papers, University of London

and Lubbock, which was a pretty cool rejoinder considering what had actually occurred, but said that he "had approved the other day a book in the German series" for publication in Britain and America (probably Bernstein's *The Five Senses of Man*), and had told King "it was hardly worth while to bother you & Tyndall about it".[35] He continued: "there was a bosky French book about pleasure and pain which I positively said should have your verdict & another in French which I referred to Tyndall". In other words, he was making decisions about what Spencer and Tyndall should see and what they should not see. This brings us closer to the problems of coping with someone of Huxley's ilk. Pathologically, Huxley within himself would always find it difficult to accept that his judgement, on any subject would not invariably be superior to that of others, though he sometimes masked this by going through the motions of consultation, sincerely believing himself to be sincere.

During the same year, the tensions between publisher and committee, and between the individual members of the committee, were heightened by King's decision to advertise forty forthcoming books that had probably not been submitted to the committee's scrutiny. He had had enough. He was now explicitly at odds with the committee because its officious intrusions, as represented by Huxley, were negative, obstructive and unhelpful. The decision to advertise forthcoming works reflected his perception as publisher that the Series was doing well. Because he remained faithful to the original plan for the Series by which successful books would in England be reprinted in sets of 1250 copies, he was in a position to know how the Series was developing, as a series, which the self-appointed committee was not. Despite the amount of print-shop activity needed to keep the Series going, King had become optimistic about it, sufficiently so for him to advertise forty new books as forthcoming. In whatever manner the committee chose to define its role, the fact of the matter was that the publisher was necessarily the person with whom an author had dealings. It was the

[35] Huxley to Spencer

publisher who received unsolicited proposals for books from British authors, just as it was usually the publisher who received manuscripts, books and queries from French, German and Italian publishers. There are not many publishers, if there are any at all, willing to take all the financial risks associated with any publication while other people make the decisions. It was the publisher's name that appeared on the title page, not the names of advisers or committee members. However much a publisher might welcome advice, and indeed quite frequently seek it, there would always be a point at which he had to step back and make his own final decision to publish or not to publish. This fact of life was unacceptable to Huxley. In any enterprise Huxley always wanted control, particularly, it seems, in this one, so to him it mattered not that he knew very little about publishing, let alone publishing as a business, since he had convinced himself that he alone should be the arbiter of what constituted good science and what constituted bad.

The matter had come to a head when Huxley again acted to curb the independence of the publisher, requiring that a new contract should be drawn up between the publisher and the representatives of the authors, which would clarify the role of each party in a manner not achieved, Huxley thought, in the St. James Hotel document. Such was his animosity towards King at this point that he completely forgot, or chose to overlook the fact that the International Scientific Series was an Anglo-American project, that Youmans had been its principal pioneer, and that the scheme existed to facilitate simultaneous publication in Britain and the United States. In October that year Huxley called on King at his Cornhill office to thrash out the matters he believed required clarification: the authors' remuneration and the powers of the committee. A new contract was drawn up and, after further discussion, was amended to specify that, while academic decisions would be made by the committee, commercial responsibility would continue to rest with the publisher. This was one of Huxley's many Pyrrhic victories. He derived great personal satisfaction from imagining that others were against him, forcing him against his better nature to enter into combat so that common-sense and logic would prevail. It was easier for him, psychologically, if the publisher was an enemy. In this particular case, however, it is not certain that he did prevail. Though the *draft* contract has been preserved among the Huxley papers at Imperial College, there is no evidence that any actual contract was ever signed. It would hardly be surprising if King, having made such an effort to get the Series going, took umbrage at the suggestion that a new contract was needed. At all events, what he believed he had gained was the provision that authors would be paid £55 when a book first appeared instead of £50 as originally agreed.

Huxley told Tyndall by letter that King had been "surprisingly reasonable" and was quite willing to enter into a new agreement "with us". Characteristically, he said in the same letter that he had agreed to King's proposal, though what he meant was that he had made a proposal to which he believed King had assented. He said he had acted as Tyndall's "ambassador" because the two men had discussed the decision-making problem at some earlier point in time, but by this stage Tyndall was not particularly concerned either about the money or the power of decision. The negative reception of his own book had not increased his interest in the day by day workings of the Series. Herbert Spencer was not mentioned in this Huxley-Tyndall correspondence. He was, though, very concerned about the turn of events, realising that the publisher had little reason to keep the Series going if obstacles were perpetually being thrown into his path, so he wrote to Youmans suggesting he might wish to return to Britain to rescue the Series. If Fiske is to be trusted, Youmans responded to Spencer's letter by travelling to Britain in June 1875, but he was unable to achieve the reconciliation between King and Huxley needed to keep the Series going in the manner originally planned. Consequently, Herbert Spencer stepped down from the committee, leaving Huxley as the programme's only "adviser", a position of dubious utility he would hang on to for several more years. Sometime later, Huxley tried to explain away "the ferocious onslaught we made upon you", admitting "I have not a few times been ashamed of the extent to which I have tired your patience".[36] The onslaught was nonetheless terminal as far as their relationship was concerned, and also as far as the International Scientific Series was concerned. The advisory committee, as a committee, no longer existed, though Spencer continued to support it in all the countries involved except Britain.

A fairly clear indication of the damage done to the British part of the Series by the in-fighting that Huxley seemed so much to enjoy is provided by a glance at how the Series was doing in the other countries involved. Thus by 1877 there had been two printings of Bagehot's book in Paris, two of Bain, Stewart, Maudsley and Jevons, three of Herbert Spencer and four of Draper (American but obtained by Baillière through London). In as far as these authors were doing rather well in Paris, there was no reason for the French publisher to have doubts about his business alliance with King since it seemed that the enterprise would be a mutually profitable one, but King did not publish three of the works that appeared in Paris—those by Dumont (1875), Bertholet (1876) and Brialmont (1876). Similarly, though the Series got off to a flying start in Leipzig, several of the early volumes were not translated into English for co-publication in Britain; Fuchs (1875), Peters (1876), Brucke (1877) and Hugo von Meyer (1877). Then in the following years, when the effects of the London crisis were still being felt, Bohmert (1878) and Hartmann (1879) were also passed over in London. Of works published in Milan, Blaserna's *Theory of Sound* appeared in London in 1876, but the next three Italian works were not published by King. Though these figures have to be seen in the context of King's overall publishing programme in which five German, two French and one Italian works *were* translated and published in London and New York, the rejections tend to confirm there had been a severe reversal of attitude in London, where a series of popular scientific texts

[36] ALS: Huxley to Spencer. Huxley Papers at Imperial College, folios 99 and 100, dated 28 November 1876

was apparently still wanted but where the spirit of international collaboration had been snuffed out.

Huxley by no means accepted any blame for this change of direction. On the contrary, the crisis had confirmed his own convictions. As usual, he thought his opinion was the only logical one that could be held. He genuinely believed, or at least behaved as though he believed he was right, and, also, that London was the centre of the universe as far as science publishing was concerned. He did not pretend to have deeply felt international sympathies. His involvement in the work of the Series had revealed a strong tendency on his part to suspect that scientific work conducted in other countries would be inferior. A different person might say: suppose it is in some sense inferior, it might still be worthy of publication; let the reader decide. This was not Huxley's way. He was always absolutely sure that he, uniquely, was the one in a position to decide. Then there was the question of the readership at which the Series was aimed and therefore the level at which the discussion of scientific subjects should be pitched. Throughout the early part of his career, and probably for the rest of his life as well, Huxley consistently thought that popular science should be at the lowest possible level, that there existed readers who would know very little and understand less, and that the popular science writer should therefore write in as simple a manner as possible, avoiding complexity, technical language and references away from the text. While the International Scientific Series was being launched, he had accepted the position as general editor of a science series being published by Macmillan, without feeling that there was a conflict of interest or loyalty, and for Macmillan had been able to implement his belief that introductory science primers, intended to introduce subjects utterly new to non-scientists, should be set at a low level where unreasonable demands were never made on the uninitiated reader. The Macmillan series was successful in this respect. He therefore believed the International Scientific Series should be run in exactly the same way. As things turned out, some of his contemporaries agreed with him while many did not.

Among those who did not were talented scientific authors who, in a variety of ways, faced up to the challenge of introducing complex, specialised subjects to readers intelligent and determined enough to accept the challenge and profit from it.

If one looks back at the events of the early eighteen seventies which have been briefly summarized in the preceding pages, it becomes clear that in the English part of the Series its original mandate had been forgotten, at least by Huxley. Appleton, Youmans, Spencer and to some extent Tyndall, together with their advisers and consultants on both sides of the Atlantic, had optimistically believed there were books to be written, sold and shared, which celebrated significant "advances" in modern thought, that these advances were a function of thinking scientifically, that both technology and the social sciences constituted important elements in modern thought, and that there were readers anxious to know about discoveries, inventions and intellectual developments that had a bearing on how people lived—that is, on how they themselves lived. These "advances" represented an increase of human power over the natural world, allowing people to think more confidently about both physical and social progress. Something could at last be done about the problems of sewage, drinking water, drainage and the control of disease in the Thames basin. Better ships with improved sources of power could be built on the basis of an increased understanding of the engineering problems involved. Tunnels could be constructed under the Thames. Medical care could be improved. And so on. The story of the nineteenth-century transformation of Britain into an industrial state based on advanced ideas, increased resources and greater inventiveness is a familiar one. Retrospectively, then, it is easy to see the International Scientific Series was aimed at readers interested in the way greater control over the physical environment was being achieved, readers therefore not opposed to change and progress. The Series was self-consciously modern. It did not have as its purpose either the education of the as yet totally uneducated or the repeated rebuttal of religious faith and conviction, both of these concerns remaining uppermost in Huxley's mind, in fact dominating his response not only to science but also to people.

Huxley found it convenient to forget the original purpose of the Series but Herbert Spencer did not. When, a little later, he gave evidence before the Royal Commission on Copyright, in March 1877, which of course was after Huxley had been to North America, he explained both the technological and commercial benefits of the Series in the terms originally agreed, while, arguing for an extension of copyright protection to thirty years after the author's death, stressed the advantages of Series to authors and their dependents in the long-term. However, he had withdrawn from the Series at an earlier date. While it had been functioning as originally intended, his own extremely successful *The Study of Sociology* had appeared in London and New York. 8750 copies had been sold in Britain and the revised version of the book was still in print. During the same period Appleton had put out five differently dated "editions". The book was published in Paris in 1874 and Leipzig in 1875 in two volumes. But that was the point at which he withdrew. Spencer's next book was *Les bases de la morale evolutionniste* published in 1880 in Paris but not in London. He had become part of the opposition Huxley needed to eliminate if he were to have full control of this particular publishing scheme.

The Crayfish

The purpose of this short chapter on Huxley's relationship with the American publishing firm of D. Appleton & Co. has been to show how, notwithstanding all the administrative ups and downs and conflicts and disagreements, the British scientist nonetheless developed a non-trivial bond with an American readership, so that his work as published in printed books quickly came to be as well-known in the United States as in Britain. This growing mutual awareness lay behind his decision to cross the Atlantic. The question is where to stop. This is not the place for an extended discussion of the International Scientific Series *after* Charles Kegan Paul replaced Henry King as the British publisher and *after* Huxley had been to the United States in 1876. The long and complex story of what happened next to the International Scientific Series will have to be told in a different book. Nevertheless, it is tempting to pause for a short while to look at Huxley's contribution to the Series, which was a book called *The Crayfish* published by Kegan Paul in 1880, though conceived, planned and paid for at an earlier date. In the early eighteen seventies Huxley had been paid an advance of £100 for a contribution to the Series he had helped found. Charles Kegan Paul said he wanted either the promised book or his money back.

The composition and publication of *The Crayfish* provides a cameo image of the way Huxley often worked. One step in the process was a talk he gave to the Zoological Society which was published as "On the Classification and Distribution of the Crayfish" in *Proceedings of the Zoological Society* in 1878. Based on earlier work, this was an extempore performance recorded in shorthand by a secretary employed for the purpose and then edited by T.J. Parker, Huxley's demonstrator, Huxley paying £6.6.0 for the shorthand record and £50 for the editing. Parker probably also helped with the construction of a new text to be used in the second half of the book that became *The Crayfish*. It is not being suggested that there was anything wrong with this procedure. Some scholars and scientists prefer to take total responsibility for what they write, attending themselves to all the associated chores, while others routinely accept the assistance of demonstrators, research associates, family members and hired help. Huxley was by no means alone in seeking, accepting and paying for such help, though the practice perhaps results in a noticeable stylistic unevenness in his case. On the other hand, even if he was doing what others also did, it is certain that he was intent on cobbling a book together from material he happened to have at hand, just as a wood-turner, for example, will often try to think of something to do with interesting bits of surplus wood lying about in his workshop. He had no thought of returning the £100 advance.

The footnotes provide a clue to the way the book was put together. In the earlier part of the book the footnote material pre-dates the actual publication by thirty or forty years. These notes were not revised. It had been early on in his career that Huxley had dissected, measured, analysed and described the crustacea, so perhaps reasonably enough many of the footnotes hail from that time, though possibly Parker up-dated the references to Bell's *British Crustacea* when he edited the book. Be that as it may, the references to Réamur, *The Magazine of Natural History* and *Annals and Magazine of Natural History* look more like intellectual fossils than as belonging to the eighteen seventies. They provide one of several hints that Huxley's work on this subject had been done several decades earlier and had not been kept up to date. The first part of the book contains many such fossils as, for example, the retention of the word "race" to mean a variety within a species. Had the text been revised by the author this word would surely have been abandoned. Generally speaking, the book's organization, language, apparatus and treatment of the subject together show that it was written before 1859 and then adjusted only slightly, if at all, to the quite different research situation that prevailed in 1879. No wonder he could be curtly dismissive of evolution. The first part of *The Crayfish* was written before the importance of *The Origin of Species* was understood.

As for language and treatment, the reader will quickly see that the text shows Huxley having difficulty with a number of key problems, which remain unresolved. These include what he calls the "nucleated cell" (pp.192-222), the definition of "species" (pp.291ff.) the discussion of "transformism" (pp.317ff) and the problems of global distribution (chapter 6) where most of the information is second-hand but not acknowledged. The reader will also see that some of Huxley's problems derive from his mechanistic treatment of anatomy, as indeed has often been pointed out. The crayfish is an "alimentary machine" (p.69); the heart is a "pumping and distributing machine" (p.69); the "body may be regarded as a factory, provided with various bits of machinery" (p.84); the whole animal is "a piece of mechanism". Presumably Huxley judged these terms to be helpful to the intended readership of the Series, or they are lecture room expressions which would provide the first steps in discussions with students.

In order to write *The Crayfish* Huxley resorted to one of his favourite stratagems. In the preface he says: "I have desired in fact to show how the careful study of one of the commonest and most insignificant of animals, leads us, step by step, from everyday knowledge to the widest generalizations and the most difficult problems of zoology; and, indeed, of biological science in general".[37] He has already disclaimed any intention of writing an up-to-date scientific monograph which "would require the devotion of many years of patient study", so it is clear from the beginning that his aim, as usual, is educational, albeit educational within a strictly confined context. He uses the book to instruct the reader on how he believes a student should think if he wishes to be a zoologist, a subject on which he, Huxley, has strict, perhaps somewhat restrictive opinions. This educational approach, if such it is, contrasts sharply with genuinely scientific work such as George Rolleston's *Forms of Animal Life* (Oxford: Clarendon Press, 1870). Rolleston, a one-time pupil of Huxley's, sets himself the task of introducing students to problems of scientific classification which he says should be within a student's powers after a "comparatively small amount of previous preparation", provided that the hands-on dissection of specimens is a routine part of the study of all mammals, and that the student's experience of dissection always precedes the discussion of classification. From this it follows that Rolleston's description of the *Common Crayfish* (pp.204-210) will be markedly different from Huxley's in *The Crayfish*, though Rolleston says on p.210 "for a full account of the anatomy of the crayfish see Huxley, *Medical Times and*

---

[37] *The Crayfish An Introduction to the Study of Zoology,* The International Scientific Series Vol. XXVIII (London: Kegan Paul, 1880), pp. [v]-vi

Gazette, Feb. 7, 1857, et seqq." This comparison helps to clarify the matter. Rolleston in 1870 was publishing a scientific work designed to be useful to first year college students. Huxley in 1880 was publishing an introduction to the same subject designed to be accessible to a wider, less well-educated readership. Huxley cannot be faulted for not wishing to write the kind of book Rolleston had already written, but it is clear that by the eighteen seventies his interest had shifted away from scientific enquiry, as friends like Darwin and Hooker fairly frequently noted, and had re-located itself in the type of scientific popularization that could be pitted against religious belief. He had chosen, in the eighteen seventies, to be less of a scientist and more of a controversialist.

As mentioned above, Huxley claimed that in *The Crayfish* the reader would be guided from the careful observation of a common object to wider generalization and more difficult problems. Does this happen? In the concluding chapter of the book (Chapter VI) Huxley resorts to what might be called the Whitehall negative on at least twenty five occasions within the space of fifty eight pages (pp. 288-346). These negatives take the syntactical form of "I cannot hear of any …" (288), "it is impossible to say" (295), "it is impossible to decide" (296), "it seems not improbable …" (321), "the question is involved in some obscurity" (289), "so far as is at present known" (305) and so on. There is a cloud of explicit uncertainty over the whole chapter. Far be it for a non-scientist to suppose Huxley is admitting ignorance of his own subject, though some of his statements are alarmingly cavalier and imprecise as "I have no doubt they derived from ancestors which lived together in the sea" and "I think that all that can be safely said, at present, is that the existence of intermediate forms is not proven". When writing this final chapter, Huxley appears to have forgotten that he promised to conduct the reader from observable fact to securely tenable generalization. Was Huxley personally responsible for the wording of this chapter or did he take on trust the version that his assistant transcribed from shorthand notes? It looks as though he either did not read it at all or else did not read it with sufficient care.

That he wrote the final page is, however sufficiently clear. The book ends with these two final paragraphs.

> Thus, with respect to the aetiology of the Cray fishes, all the known facts are in harmony with the requirements of the hypothesis that they have been gradually evolved in the course of the Mesozoic and subsequent epochs of the world's history from primitive Astacomorphous form.
>
> And it is well to reflect that the only alternative supposition is, that these numerous successive and coexistent forms of insignificant animals, the differences of which require careful study for their discrimination, have been separately and independently fabricated, and put into the localities in which we find them. By whatever fog the question at issue may be hidden, this is the real nature of the dilemma presented to us not only by the crayfish, but by every animal and every plant; from man to the humblest animalcule; and from the spreading beech and towering pine to the *MIcrococci* which lie at the limit of microscopic visibility.[38]

This formulation *is* characteristic of Huxley, whether he wrote the words on a piece of paper himself, or someone did it for him. The word "dilemma", as used here, is mere rhetoric. It is quintessential Huxley. He cannot help himself. The reader of the book is not to be imagined as believing the global distribution of varieties of crayfish within the same species represents the hand of any god, Christian or otherwise, in separate acts of creation. He or she, if religious, will obviously have thought long and hard about how to reconcile genesis stories with modern science, so it is reasonable to ask, in retrospect, whether Huxley was serving his cause well by indulging in these simplistic generalisations. The artifice is actually rather unseemly, in as far as the reader knows Huxley does not think a dilemma genuinely exists.

To return, however, to the main line of this account, leaving individual readers to decide for themselves what the merits of *The Crayfish* are or are not, Huxley certainly made heavy weather of the eighteen seventies as he agonised

---

[38] ibid. p.346

over the direction his career should take. His wrangles over the International Scientific Series had exposed various psychological problems. Was he always right? Could he get on with other people or did he always want to dominate them? Had the time come for him decide between a full commitment to a research career or, for the rest of his life, did he prefer to devote his best energies to controversy and debate? And could he free himself from his debilitating preoccupation with other people's religious beliefs? These questions might be answered one way or another by his taking a long break from his London routine. His journey to America would be crucial.

CHAPTER TWO

It is not surprising that Huxley's five week visit to the United States has already been described by several authors, since it was certainly one of the more significant events of his life. His son, Leonard Huxley, wrote a kindly, low-key account of it in Volume One of his *Life and Letters of Thomas Henry Huxley*, celebrating the journey his parents had made together—the first together for twenty years, he said—and treating it, not as a crucially important turning point in a scientific career, but as a natural and logical stage in the establishment of the international reputation he eventually acquired. Many years later, in Volume Two of his *Huxley*, Adrian Desmond wrote an extremely useful, twenty-page chapter about the same visit, having at his disposal the research results of his characteristically thorough enquiries. In between these landmark publications there have been other biographical essays, the more compelling of which are listed in the bibliography of the present work.

When what is at stake is the historical recollection of a human being of some importance, that is, someone who left his or her mark on a culture as it existed at a particular time, it is not surprising that, looking back, biographers are struck by different aspects of their subject, to some extent responding to the facts, as they see them, that their research reveals, and to some extent honestly interpreting these facts to the best of their ability, though under the influence of their own predispositions, values, attitudes and experience. It soon becomes obvious that there cannot ever be a definitive biography of any historical figure, however well known, which means, if true, it is always better that there should be several, or many biographies, rather than one. Where there is only a single biography of an important historical figure, the interested reader or scholar is in the same position as a tenant, servant or woman on a large, eighteenth-century European estate who could only know what the property owner allowed them to know, which in many cases was not very much, sometimes, as subsequently described and underscored in novels by George Meredith, for example, being only allowed to travel in proscribed directions, literally not knowing that there were other places of interest that might have been visited. Over the years there have certainly been biographers who have behaved like old-fashioned property owners, but that is not the intention in the present book.

A fully satisfactory biography of Huxley has still to be written. There are many unresolved puzzles still to be pondered. For instance, he wore spectacles when reading, but preferred not to use them when lecturing. Several images show him wearing spectacles, as for example the reproduction of the pencil drawing by Wiroman reproduced in the Huxley website. He did not have the sort of glasses that would allow him to read a script *and* see his audience. His lectures were therefore seldom written out in full, if at all. The text of any lecture or address invariably had one form and one set of detail when delivered and a slightly, or greatly, altered form when reconsidered, at which point it was usually being written for the first time by Huxley in his study at home, though he would have the help, at home, of notes first made in the lecture room by a shorthand typist or secretary. This was the case with all the lectures delivered in America and subsequently prepared for publication in England. Perhaps partially as a consequence of his eyesight problem, he preferred, in nearly all his lectures, to develop his subject by deliberate steps from a simple beginning, avoiding both syntactical complexity and complicated subjects of the kind he considered intellectually too demanding for a particular audience. A consequence of his doing it this way was that he mostly gave a lecture a straightforward, linear, "logical" structure, without digressions or anecdotes, which many found helpful, but some found simplistic and reductionist. Was this habit or preference his way of coping with a personal difficulty? And do his lectures as published bear the marks of a difficulty that was never resolved? Was his mostly unscripted lecture-room rhetoric an advantage or a disadvantage? It needs to be said that nothing in this paragraph is intended as a criticism. The purpose is no more than to draw attention to his preferred mode of functioning, which, though not unique, was unusual.

Huxley's smoking raises a similar question. He was a heavy user, or abuser, of pipe tobacco according to his son, and many others besides. As he grew older, he resorted more and more to this particular pleasure or solace. Was his smoking related to his digestive problems, his ill-health generally and, for that matter, to the way he thought? Might he have thought differently about his own life experiences had his small study at home not been permanently contaminated by tobacco smoke? Huxley usually described himself as a comparative anatomist, not as a physiologist. He spent at least as much time thinking about skeletons as about living, flesh and blood creatures, a bent that was partly determined simply by what could be seen in museums and laboratories at the time, but was also partly a consequence of his cabinet life-style. To keep up with the field he needed time—a lot of time—in his study and when there he liked smoking. Though the elusive question of Huxley's long-term ill-health has so far remained unresolved, his over-dependence on tobacco is one of the factors that needs to be considered as possibly relevant.

Another factor that militates against the writing of a straightforward narrative account is the Victorian preference for keeping the public completely separate from the private. Huxley was unlikely to leave a record of his personal feelings and thoughts when he and his wife travelled to Liverpool to embark on their trans-Atlantic voyage. What kind of cabin they had, whether he liked it or not, what he thought he was doing in leaving London for such a long period, whether he smoked in the room where they slept, whether they dined with travellers they found interesting, whether they struck up lasting friendships—these are matters not likely to be made openly revealed by the stereotypic kind of middle-class Victorian Huxley had chosen to become. A man's private life was nobody else's business, an attitude gently satirized by Dickens in *Great Expectations* where Wemmick's suburban home is represented as a castle with drawbridge and battlements. "An Englishman's house is his castle" was a saying that had already gained currency. In Huxley's case the effect of this perfectly legitimate social preference has been to de-humanize him somewhat, as though only his public life was, and is now, important and as though his private habits, patterns of thought, preferences, likes and dislikes and fallibilities never had a bearing on how he conducted himself in the public domain of work, career aspiration and achievement. His wife sometimes allowed herself to comment on this, telling friends that if they saw her husband at home they would realize that, even if he was ruthlessly competitive in the public arena, he also had a kindlier, more good-humoured side that made him an engaging companion when out of the limelight. At all events, there are good reasons for hoping there will be other biographies of Huxley in the future where less is taken for granted, as much is currently taken for granted in the received story of his life. For the moment it looks as though he denied himself many of life's rewarding pleasures for the sake of a successful, remunerative public career. He seems never to have had much time for art, music and literature, for example, though further speculation on this subject will be delayed until the next chapter since it is time to get him across the Atlantic from Liverpool to New York.

The five main parts of this chapter will be about his week's work in New Haven at Yale, his journey westwards as far as Niagara together with at least one virtually impromptu speech in Buffalo (25 August), his foray to the south as far as Nashville, Tennessee, to see his sister, where he gave at least one off-the-cuff speech (7 September), his inaugural address at the Johns Hopkins University (12 September) and his three lectures in New York (18, 20 and 22 September).

It is fairly certain that most of this had been arranged in advance, even though the planning operation cannot be fully documented. As anyone knows who has family on a distant continent, the family motive can be extremely strong, so to a certain extent the trip may well have been built around the desire for reunion, however brief it had to be. There was also the perennial health problem.

Supposing that a real break from his punishing London routine had been judged necessary, and that a trans-Atlantic voyage had itself been thought possibly good for him, it was of course also the case that he would wish to derive maximum benefit from his travels, the people he met and the new experiences he could expect to have, as indeed turned out to be the case.

Huxley's publications in North America

When the "*Germanic*" reached New York Huxley and his wife were met by his publisher who took them, either immediately or the next day, to the Appleton residence on the Hudson River. There was already a strong rapport between publisher and author, based partly on their meeting in London, partly on the energetic spade-work of Youmans, but chiefly on the publishing programme itself. The founder of the company, Daniel Appleton, had re-located in New York in 1825 where he published his first book in 1831. When Daniel died in 1849, the business passed to several members of the family of whom William turned out to be the most active, especially in the business of attracting European authors to the firm as it continued to prosper.

By 1876 Appleton already had eight Huxley titles on his list. These were *On Our Knowledge of the Causes of the Phenomena of Organic Nature* (1863), *On the Origin of Species* (1863), *Evidence as to Man's Place in Nature* (1863), *Elements of Physiology and Hygiene* (with E.L.Youmans,1867), *Lay Sermons, Addresses and Reviews* (1871?), *Manual of the Anatomy of Vertebrated Animals* (1871), *More Criticisms on Darwin, and Administrative Nihilism* (1872), and *Critiques and Addresses* (1873). The selling price of these affordable cloth-bound volumes was between fifty cents and two dollars fifty (for the *Manual of the Anatomy of Vertebrated Animals*). As already mentioned, Appleton could reach a much larger readership in America than Macmillan could in Britain while at the same time, though the religious or clerical resistance to Huxley was strong, it was not as strong as in Britain where the Church of England continued to wield excessive political power, as did the Evangelical Movement in its dying phases. That Appleton was not bothered by the subsequent ups and downs of the Huxley American visit is clearly indicated by his continuing to publish Huxley's works for many years, specifically *The Crayfish, Science Primers: Introductory, Manual of the Anatomy of Invertebrated Animals, American Addresses, Physiography* and *Essays on Some Controverted Questions*. (The details are given in the bibliography). Though the Appleton list is certainly impressive, and indeed suggests that Huxley was at least as well known in America as in Europe, the reception of Huxley in the United States obviously did not depend only on the reading of the works listed above, since many Americans had access to British and other European journals and magazines. The only Appleton-Huxley publishing contract that has been seen to date, that for *Essays on Some Controverted Questions* in the Lilly Library at the University of Indiana, guaranteed Huxley 10% on the retail price of copies sold, with provision for what would happen when the book stopped being profitable, but in fact all the books listed above were still in print in America when their author died. Huxley and William Appleton sustained a friendly business relationship for the rest of Huxley's life, meeting for the last time in 1892.

When Appleton published the book called *American Addresses* which will be discussed later in this chapter, he had included an eight page list of scientific works published by the firm, almost all of them by British scientists and social scientists, including five by Alexander Bain, three by Charlton Bastian, six by Darwin, ten by Herbert Spencer and six by John Tyndall. As already noticed, the International Scientific Series was simultaneously part of the Appleton publishing programme. British scientists made up the whole of this part of the Appleton list, with only one or two exceptions. He had found a commercially viable niche within which the firm, or its scientific arm, continued to prosper at least up to the First World War. Some publications were obviously less important than others. Huxley helped Sarah Blackett with her book with the disarmingly ambiguous title of *Boys and Girls in Biology: or Simple Studies of the Lower Forms of Life.* (Though Huxley may have thought that boys and girls were low forms of life, this is not what the title means). He seems not to have contributed to the anonymous *Tommy Try, and What He Did in Science,* also advertised in the same list. By and large, however, William Appleton did rather well from late nineteenth-century British science writing.

A "Sketch of Professor Huxley" by Ernst Haeckel, part of which had previously been published in *Nature*, appeared in Appleton's *Popular Science Monthly* in the November 1873-April 1874 issue, almost certainly at the instigation of E.L. Youmans who wrote the final part of it. The last paragraph reads as follows;

> No statement of the character and work of Professor Huxley would be complete that did not recognize his remarkable attainments as a writer. All who have read the mastery papers contained in *Lay Sermons* or the *Critiques and Addresses* will acknowledge his fine and vigorous command of English, and the literary richness of his style. He has a keen enjoyment of literary excellence, and "keeps up" with poetry, fiction and the progress of critical thought, notwithstanding his indefatigable scientific investigations. Owing to these traits, Prof. Huxley has a high reputation as a popular scientific teacher; and even his *Lectures to Working Men* are models of what such discourses should be—clear, simple and attractive, yet carefully accurate and strictly scientific. As a public speaker he is quiet, deliberate, fluent, and, we might almost say, colloquial; while socially he is genial, witty and brilliant. He is, moreover, a man of enlarged sympathies, in this respect contrasting markedly with many scientific men, who are swallowed up in their specialities, and never give a thought to anything beyond them. Prof. Huxley has, however, overworked himself, and damaged his health. We hope he will regain his power, and be able to give this country a session of lecturing as he has long hoped to do. He is perhaps the only scientific man in England who could revive for us the pleasant experience we had with Prof. Tyndall last year.[39]

[39] Appleton's *Popular Science Monthly*, Nov. 1873-April 1874, pp.739-743

This eulogy has the ring of sincere, friendly conviction. Though Youmans knew about Huxley's disagreements with Henry King, he probably more trusted the scientist than the publisher, while, having met Huxley socially, knew he could be good company. It has to be doubted that Huxley read very many novels, which was a form he did not understand, did not have time for and tended to deprecate as not having anything to do with the rational conduct of sensible individuals in public life. When J.B. Hooker wrote to Huxley to say he had just wasted a rainy Sunday afternoon reading Richardson's *Clarissa*, he evidently knew Huxley would agree with him. When Huxley could have been reading the most recent fiction as he crossed the Atlantic, he was probably reading the works of Hume, having had to decide what books, if any, he could sensibly add to his luggage. His own book on Hume was published just a few years later; the voyages would have given him time to reflect on a subject with which he was already very familiar. If Hume travelled with him across the Atlantic, as seems probable, it would have been in the pocket-sized, four volume edition of 1826, first published by Black & Tait in 1828, which had been in Huxley's possession for many years and which he mentions on pp. 45-47 of his *Hume*. However, the subject of Huxley's books, the ones he had in his study at home, has not yet been addressed in a thoroughgoing manner, so caution is needed. No doubt the Huxley home was full of books suitable for Huxley's wife and the children, which Youmans would also have seen.

The Visit to Yale

Take wheat straw, for example, as an indicator of long-term cultural stability. In Huxley's day the concept of genetic modification was far in the future. Straw was a known commodity, it varied only according to the conditions of production, and its uses were important but limited. In England, it was used as bedding for cattle, for thatching cottage roofs, for re-enforcing and insulating floors, for packaging crates and barrels for the transport of fragile or valuable articles, for stuffing mattresses, and for making effigies and other such decorations and abominations. Probably there were uses not listed here; the point is it was a reliable, stable product that was dried in the fields in stooks and stored in stack-yards by folk who knew how to make a stack that would stay dry for several months. The idea that you might want shorter straw, or wheat that needed a shorter growing period, or wheat flour with radically different properties, remained unknown. The scientific future of agriculture remained unknown. How could you consider something that had never happened, that was way beyond your ken, that none of your friends and neighbours had ever experienced or thought about? And looking at the matter from a particular location—the place where you happened to live—why would you want straw to be different from what you knew it to be like? Conceptual stability was much easier to live with than instability. This was Huxley's position in 1876. It was easier to assume stability in Nature than to think otherwise. Concepts of instability were too difficult for most Victorians.

When the swallows returned to the outbuildings and gables every year, they were constant in size, structure and colour. Certainly you had been told, or had actually observed, that swallows in other countries might be larger or smaller according to the conditions of their existence, but you never saw a red swallow, or a giant swallow, or a bird with four wings. Nature was something you could reliably know about. Knowledgeable naturalists could write field-guides to help you distinguish between birds that were unfamiliar or, a first glance, identical. Poets could write about "nature" as though the experience of the external world was constant and universal. There was no ingredient of doubt in the use of the term "universal". Deviations from the biological or physiological norm could be dismissed as "freaks", "throw-backs", "runts" or the accidents of the birth process. Huxley grew up into a world in which nature seemed to be constantly repeating itself. If you killed a goose for Christmas it would taste the same as the goose you enjoyed the previous year and would never be mistaken by a blind person as roast water-hen or ostrich casserole. This being the case, it is not difficult to understand why Huxley, like most of his contemporaries, had problems when it came to thinking about biological modification over time. Besides, where was the evidence? The evidence, if it existed, still had to be collected, checked and analysed..

Huxley was not psychologically astute, imaginative or inventive. He preferred to tackle plain, definable problems in a plain, ostensibly logical way. There are many unpublished manuscript fragments in the Library of Imperial College that demonstrate the methodical manner in which he tried to come to grips with a problem that puzzled him. For example, when the fossil skeleton of a previously unknown creature was found in a place called Cummingstone in the north of Scotland, he wondered whether it might have been a crocodile. The note at Imperial shows him checking the anatomical features of the fossil skeleton against what he knew about crocodiles. It was easier for him to think about something he knew than something he did not know. But crocodiles in the north of Scotland!? It was a puzzle he was unable to solve, since he had no knowledge of plate tectonics or the movement of land masses over long periods of time. Putting the matter very simply in the manner of which he would have approved, it was next to impossible, at least before his American adventure, for this medically-trained anatomist to conceive that what he knew as the skeleton of any creature—a skeleton that he could measure and weigh—had undergone physical changes during long periods of time, which was the scientific conundrum he frequently dwelt on as his reason for not giving the Darwinian hypothesis his full assent. It is because this was his stumbling block that *Man's Place in Nature* is so important for anyone wishing to understand his intellectual predicament. He would only be satisfied, that is, he would only be able to satisfy himself, if someone showed him a continuous series of anatomical variations or apparent anatomical developments from roughly the same part of the world. He had learnt that Professor O.M. Marsh at Yale College might be able to do this, or rather he had read Marsh's short 1874 article about fossil horses in *The American*

Naturalist. Visiting Professor Marsh in New Haven therefore became an important item in his travel agenda as he had explained to his friend Moncure Conway. "One of my objects in going to the States is to see Marsh's great horse", he had told him in April.[40] The two men were already acquainted, Marsh having made a point of meeting Huxley during a visit to Britain several years earlier.

The part of Professor Marsh's research that Huxley knew most about was his work on the skeleton of the horse, an animal which, if not truly indigenous, seemed so at the time. It certainly turned out that there had been wild horses in North America over a very long period. The age of their skeletal remains could be determined geologically according to the type of rock in which they were found, that is, once the geological survey work had been carried out. Their presence had not been known to early nineteenth-century travellers and explorers, or had not been reported, when the western territories had yet to become states and travel was still easiest by canoe, though still fraught with all manner of danger. The first federally employed geologist had been George Featherstonhaugh whose exploration of the western territories during the eighteen thirties (but still east of the Rockies) were recorded in his reports to Congress and two books first published in Britain. But "Feathers" had to travel by canoe, on horseback or on foot. As is well-known, it was the railroad that made the difference. Unlike their predecessors, scientists were able to travel to the uncharted, to a large extent still unexplored West by train and, by train, return to Philadelphia and New Haven with their finds and field results. Specifically, Marsh and his colleagues were able to use the Northern Pacific Railway, which had been planned in the eighteen sixties, begun in 1871 and eventually, in 1883, would connect Chicago and the Great Lakes to the West coast along a route just to the south of the 49th parallel. With headquarters first in Brainerd, Minnesota, and then in Saint Paul, the Northern Pacific was the first railway to connect all the

---

[40] ALS: Huxley to Moncure Conway, 18 October 1876, Butler Library, Columbia University. Among the other Huxley letters in the Butler Library is Huxley's to Conway dated 18 April 1876

northern States west of the Great Lakes and, with branch lines, Manitoba and British Columbia as well. This is an instance, then, of advanced technology facilitating academic research in a manner previously impossible. This particular railroad was built both from west to east and from east to west, but it is only the latter that is of concern here. The line was extended from Minnesota to what is now North Dakota during 1870, then in 1872 a further 164 miles of track were added across North Dakota, where the surveyors, engineers and work crews found they needed military protection from attacks by native Indians—protection which was perhaps also helpful to the scientists from New Haven and Philadelphia. Despite financial difficulties, work on the line continued until 1875 when the Northern Pacific was temporarily brought to a standstill by bankruptcy and George Custer, in the same year, was sent to Fort Rice, Dakota Territory, charged with the task of protecting the railroad from further attack, the situation having become so tense and difficult that further scientific work became impossible, at least for a good number of years. Once Custer and his men had been eliminated by representatives of the Sioux Nation the incentive to conduct further palaeontological research in Wyoming and Dakota territories was not high. Professor Marsh was never able to do further work in the territory where he and his team had made such important finds.

In the case of the horse, the finds were massive, so whole railcar loads of bones and fossils, it is said, were soon being transported eastwards by train for examination by Edward Cope in Philadelphia and Othniel Marsh in New Haven.

Othniel Marsh (1831-99) had grown up in Lockport, New York, (near Niagara), had attended the Phillips Academy at Andover, and had graduated at Yale in 1860, after which he took refuge in Europe for a while, studying anatomy and palaeontology at the universities of Berlin, Heidelberg and Breslau. Returning to America in 1866, he was appointed Professor of Vertebrate Palaeontology at Yale. With the financial support of his wealthy uncle, George

Peabody, Marsh was able to assemble a team of assistants to help him search for fossils in the Badlands of the Far West and was soon sending their finds back to New Haven. During the late eighteen seventies, and in later years, they began to find the remains of previously unknown dinosaurs, flying reptiles and toothed birds, eventually describing at least twenty seven previously unknown dinosaur genera. This work was begun in 1870, involved four or five boldly conceived, large-scale field-trips to Dakota and Nebraska, was sustained for more than thirty years of further work at home base in New Haven, and resulted in many publications, including *The Dinosaurs of North America* (1896).

Although some have found it easy to dismiss Marsh as merely an arm-chair palaeontologist, that is not how he was seen by his contemporaries. Over the years, he has sometimes been criticized for having the support of a wealthy uncle, as though being able to afford costly research trips were some kind of personal, moral deficiency, and as though his often partisan critics would not themselves have encountered difficulties of any kind as they quickly re-assembled, catalogued, analysed and described wagon-loads of fossils and bones. (Save us from that little band of arm-chair critics safely ensconced in well-paid, tenured appointments!). Marsh would later serve as the President of the National Academy of Sciences for twelve years (1883-95) being widely recognised as one of the country's leading scientists. Marsh's dinosaur research was not, however, what attracted Huxley to Yale; it was Marsh's article, *Fossil Horses in America,* published in 1874. Huxley had introduced a discussion of fossil horses into his Geological Presidential address in 1870. He now saw that much more was known about fossil horses than he had realised, specifically that Marsh had identified a succession of horse-like creatures over vast periods of time, starting with one that was no larger then a fox, progressing to one the size of a sheep, then to one the size of a mule, and eventually to a modern horse, this progression being marked not only by size but also by anatomical differences, such as the changing number of toes. These remains had been preserved in mud basins caused by the successive flooding and draining of extensive inland lakes during the Tertiary or Miocene period. In the article, the geology does not need to be precise, because Marsh's purpose is simply to draw attention to his team's discoveries. The terms "Badlands" and "mauvaises terres" are still used for the area in question. What may have particularly attracted Huxley's attention to this article in the *American Naturalist* was the following summary statement:

> The large number of equine mammals from the Tertiary deposits of this country, and their regular distribution through the sub-divisions of this formation, afford a good opportunity to ascertain the probable lineal descent.[41]

That was the opportunity Huxley intended to grasp. He spent a full working week with Marsh at Yale. He later said he had enjoyed himself immensely, as well he might, since he was being given privileged access to the raw material or data that would form the basis of later analytical work to be done by others and which he quickly realised he could use in his own lectures without, of course, admitting he had been mistaken in his previous pronouncements on the anatomy of the horse. It has often been noted that mid-nineteenth century palaeontologists were at a disadvantage whenever they found single specimens of fossils that were new to them, since it was tempting to regard any new find as a discovery that required a new species name, but this was not the case with the bones and fossils at Yale where there was much more than any individual could immediately interpret. What would Huxley make of it all?

Although Desmond labelled Marsh as "one of a new breed of robber barons", the transport of skeletons and fossil remains away from the places where they were found to locations where they could be conserved and studied was not seen as an intrinsically evil activity in the nineteenth century, and in any case

[41] O.M. Marsh in *The American Naturalist,* Vol. 8, No 5, pp. 288-94, Peabody Academy of Science, Salem, Mass.

what Marsh did in having the team's finds shipped back to New Haven was not the same as removing fossils, skeletons, artefacts and works of art away from the country of origin to an entirely different country, as has happened in parts of Africa and elsewhere. At the time, it would have been absurd for Marsh to think that what he found should have been left where he found it. When Marsh and Huxley met, it is fairly certain not much time would have been devoted to the discussion of habitat. In sharp contrast to photographs of Marsh's team on site, Huxley, though nominally a palaeontologist, had never felt the need to engage in fieldwork. He had turned up in North America in a resplendent, newly-tailored tweed suit, which was his concession to being out of London for a while, and it is unlikely that he had ever owned the work-clothes appropriate for digging objects out of the dirt. He very much preferred trying to make sense of other people's research results, was more driven by the desire to understand the processes of evolution than by any wish to learn about and appreciate the pleasure and excitement of fieldwork, and simply travelled by train from New York to New Haven to see for himself the evidence that he believed Othniel Marsh had accumulated.

Unfortunately, Huxley does not have much to say about being at Yale in 1876. The Green was of course a prominent feature, ringed in those days by a wooden fence. Connecticut Hall had been there since 1750 or thereabouts. During the early nineteenth century Schools of Medicine, Divinity and Law, as well as the Graduate School of Arts and Sciences and the Sheffield Scientific School had been added, but the undergraduate colleges that are now such a feature of the Yale campus did not yet exist. Yale would not admit women undergraduates for many years to come, though Huxley, of course, was not in the least surprised by their absence. Likewise, the impressive present-day student mix representing so many different countries of origin, and so many skin colours, creeds, religions and ethnic backgrounds, would not have been something he very much thought about. The only students Huxley had known up to this point in time were male and white. It seems likely that Huxley would have been introduced to other members of the Yale faculty, would have been invited to dinner here and there in a characteristically generous American manner, and, with his wife, would have been escorted to see the local sights, but there is scant record of such events and excursions.

It has to be doubted that Huxley's invaluable discussions with Othniel Marsh resulted in a burning desire to get on a train, westward bound, in order to see for himself the territory where the great horde of evidence had been found, since he was a palaeontologist without an appetite for field-work, and without the desire to relate finds to habitat, but impulsive enthusiasms of that kind were out of the question in any case, since he was locked into a travel programme that left little room for new plans. He was already committed to attending the meetings of the American Association for the Advancement of Science in Buffalo, and his sister was expecting to see him in Nashville after those meetings. The Huxleys would be doing a lot of travelling from this point on.

Buffalo and Niagara

From New Haven the Huxleys paid a brief visit to John Fiske whose summer quarters were in up-state Massachusetts and who had been party to Huxley's involvement with the International Scientific Series, indeed more than that, since he was one of William Appleton's more influential advisers. Fiske had spent time with the Huxleys in Britain and had been to their home, so this was a friendly reciprocal visit. From there they travelled to Boston, where they spent a night at Parker's Hotel, giving themselves time, therefore, for a brief visit to Cambridge where Huxley made a few courtesy calls on Harvard colleagues. He seems not to have known much about the Massachusetts Institute of Technology at that point, or, anyway, there is no record of his being there.

The immediate destination, though, was Buffalo, Huxley having been invited by Professor William Rogers[42] to attend the annual meeting of the American Association for the Advancement of Science., where he was somewhat lionized, his reputation in the States being already strong. This was an entirely new experience for him. In 1876, the City of Buffalo, first settled in 1789, had a population of roundabout 120,000, so the city as the visitors saw it was quite small, though by no means unimportant, since it was both a railway hub for the Grand Trunk Railroad and the western terminal of the Erie Canal and thus one of the principal, commercially important routes to the West. In the years before the Canadian ship canal was constructed, it was chiefly Buffalo's deep-water harbour that facilitated the transport of grain from west to east; Huxley would have seen the long city shoreline, the heavily laden ships and, new to him, the many grain elevators, perhaps as many as thirty. The university had been founded in 1846 but initially just as the medical school seen as crucial for the western part of the State. The present Museum of Science did not yet exist, so Huxley could not have seen its section on vertebrate evolution. Fort Porter as a military base was inaccessible to visitors and the Old Fort Niagara had yet to be restored, was in fact a ruin. The developments towards the distinguished reputation Buffalo now enjoys began a little later, that is, after Huxley's visit.

Two matters stand out from what became a busy week of networking and useful scientific liaison work. First, a set of semi-formal meetings with a group of scientists keen to see geology more fully recognized as an important discipline; second, the talk he was unexpectedly asked to give and for which he was not prepared.

The semi-formal meetings were with a group of well-established geologists, including William Rogers whose work he knew but whom he had not previously met. Although Huxley had never done any geological field-work, and never would, being best known as a comparative anatomist, he had been a

[42] ALS: Huxley to Rogers, 6 June 1876, MIT

member of the Geological Society of London for many years and had served as its President in 1869 and 1870, so was well known by reputation to the people he began to meet at the Congress of the American Association for the Advancement of Science, not least with those who were currently taking the initiative in getting geology accepted as a discipline of equal importance to the other disciplines represented at the conference.

The group Huxley joined in Buffalo was in effect a distinguished steering committee. The named members of the group were as follows.

William Barton Rogers (1804-82) had been one of the founding fathers of MIT, where he had served as its first President from 1861 to 1870, and would again from 1878 to 1881. Huxley would have known Rogers' book called *Treatise on Invertebrate Palaeontology.*

James Hall (1811-1898), by then one of the most experienced geologists in North America, had worked for many years, and with ever increasing degrees of responsibility, on the geology of New York State, after which he began to work on the thirteen volumes of *The Palaeontology of New York*, which would have been one of his principal ongoing professional concerns when Huxley met him. As one of the founder members of the International Geologic Congress, he would serve as its Vice-President at its Paris, Bologna and Berlin meetings.

A one-time student of Robert Jameson in Edinburgh, John William Dawson (1820-99) was already well-known for his field-work in Eastern Canada, notably on the fossil forests of Nova Scotia (at Joggins and at other sites around the Bay of Fundy) and, when Huxley met him, was already the Principal of McGill University, where he was also Professor of Geology. Huxley would have been aware of at least two of his books: *Air-breathers of the Coal Period* (1863) and *The Story of the Earth and Man* (1873).

J.S. Newberry was a member of the group by virtue of his work on the geology of Illinois, the fourth volume of which had been published in 1870.

Thomas Sterry Hunt (1826-92), founder member and first President of the Royal Society of Canada, had worked with the Canadian Geological Survey under William Logan until 1872, when he had been appointed Professor of Geology at MIT. Among his books were *Canada: a Geographical, Agricultural and Mineralogical Sketch* (1865) and *Chemical and Geological Essays* (1875).

Charles Henry Hitchcock (1836-), son of the famous pioneer American geologist, Edward Hitchcock, was at the time of the Buffalo conference working on the State-funded geological survey of New Hampshire which would be published as *The Geology of New Hampshire: a Report Comprising the Results of Explorations Ordered by the Legislature* (1874-78).

In September 1876, immediately after the Buffalo meetings, and after the group listed above had met a few times, Sterry Hunt addressed a notice to all geologists and everyone interested in geology—headed simply (on the copy at hand) as "Aux Géologues"—which was issued in French, German and English and was co-signed by Huxley, Otto Torell (sic) from Stockholm and E.H. de Baumhauer from Harlem in Holland. The purpose was to prepare for the International Geological Exhibition to be held in Paris the following year (as they hoped), so the announcement was that an executive committee had been formed to oversee the preparations consisting of all the American scientists listed above, with Huxley, Torrell and Baumhauer co-opted specifically to ensure that the plans were widely known well in advance.

The one-page printed document this group produced is of great interest because it proposes an agenda for international cooperation among geologists, who, up to that point, had had little option but to follow the practices and conventions of the country in which they happened to find themselves, even though it was well understood that geology was an international activity not confined within political boundaries. Even when a geologist chose to work at sites in countries other than his own he (usually still "he" in the eighteen seventies) customarily resorted to the professional, technical and academic language and specialized terminology of his own country, sometimes the language he had been required to use as a student or employee, which was then the only one he could reliably use. What the new committee hoped for was "more exact definitions, which would greatly increase the value," they thought, "of their observations and comparisons", the great collection of rocks, minerals, fossils and maps assembled in Buffalo having demonstrated the need for a greater degree of international understanding and collaboration, as well as an internationally accepted terminology. To implement this general policy the committee made three specific recommendations, with which Huxley, of course, must have agreed, since he signed the document.

It is proposed that the geological contributions to the Exhibition should include:

> (1) Collections of crystalline schists and eruptive rock, from formations identified as contact zones, as well as the effects of uplifted rock on non-crystalline altered rock. The organic remains found in crystalline strata deserve particular attention. These collections will also include every type of rock of particular importance from the point of view of chemistry, mineralogy and lithology, as well as the deposits from geysers, the variety of minerals and every kind of vein encased within rock. To the extent possible there should be, together with the rocks, preparations allowing inspection with a microscope. It would be desirable that, in the arrangement of these materials, more attention should be paid to relations in nature than to theoretical ideas or artificial classifications, so that these collections might be studied not only from a geological but also from a geognostic point of view.
>
> (2) Collections of organic remains from sedimentary rock, especially the flora and fauna belonging to horizons of special interest to geology. It appears to the committee members named below in the strata called Cambrian, ["taconique"] and ["primordial"] deserve special study.
>
> All these collections should be explained by labels, catalogues, monographs and maps.
>
> (3) Collections of geological maps, sections and models, especially sections and models calculated to throw light on the structure of mountains. In the preparation of maps it is suggested that attention should be paid to certain questions that deserve the special attention of the Congress, such as the colour codes adopted for different maps, the colours

> and symbols employed, and the best way to represent simultaneously on a single map both the superficial deposits and the strata immediately beneath them. Discussion of these questions will establish the basis for the improved design of continental geological maps.[43]

The committee's agenda for the next meeting, which was to be held in Paris two years later, was neither prescriptive nor doctrinaire but simply an attempt to indicate there were issues of methodology deserving international attention. The statement provides a useful marker in the history of geology. Thirty years earlier such a group might not have emphasised preparations for examination by microscope, for example, or the utility of models, sections and maps. Thirty years later the statement would have seemed naïve, since by then both the field-work and the publications deriving from it had become more specialized and sophisticated.

The conundrum therefore remained. On the one hand, the geologists who met in Buffalo were firmly, consciously and deliberately shifting the research emphasis away from the national towards the international, an enterprise in which, with others, they were successful, since they initiated the move towards triennial international meetings and the free exchange of information between countries. On the other hand, they seem not to have clearly anticipated the language problem. As with all scientific disciplines, the need for precision in observation, description and analysis lead to the complementary need for an appropriate language for communication between specialists—the very thing the founders of the International Scientific Series had been attempting to resist.

Huxley's position on the matter is not too easy to determine. It looks as though, in Buffalo, he may have passively acquiesced to what the other committee

[43] Only one copy of this one page, thousand word document has so far been seen, specifically the copy in the MCZ library at Harvard, which is in French and is initialled "WBR" (i.e. William B. Rogers). Because it is dated Boston 1876, there is a presumption that both Rogers and Sterry Hunt had a hand in producing it. Because a copy in English has not so far been tracked down, the translation from French into English is by the present author, though only of the committee's actual proposals (about half the document).

members wanted, since he had never had the experience of having to translate his field observations into the signs and siglas needed on an accurate, informative map. In his own publications, there are very few maps and such as there are were not made by him. When he had needed a map for the work he did in Scotland, for example, he asked George Gordon to make it for him. Generally speaking, in the second part of his career, he liked words not illustrations to do his work for him. His preference for simple, generally accessible, explicitly formulated examples and his studied avoidance of metaphor, combined with his insistence that science was more a way of life than the specialists' exclusive domain, tended to absolve him from serious consideration of the language problem. He had never been, and never would be, a field geologist. Whereas *some* of the articles he had written had been exclusively for the attention of other scientists, where there was no need to avoid words only specialists would understand, he more and more settled for a plain, non-specialist language that allowed him to address diverse types of audience on general social issues not requiring specialist scientific knowledge. By doing this he was to some extent turning his back on the scientific advances so noticeable during his lifetime, especially in the second part of it. As explained in this book and as emphasised elsewhere, he would be posthumously known, not as a scientist who had important research findings to his name, but as a successful, voluble propagandist for science, or at least for a certain type of scientific thinking. At the Buffalo meeting, however, almost everyone was English-speaking, which is perhaps why the language problem was not addressed.

As that last sentence was being written, the most recent issue of *The Geoscientist* came through the letter-box (Volume 20, No 9, September 2010), together with the sales catalogue for new and recent books. These books have titles like *The Terrestrialization Process: Modelling Complex Interactions at the Biosphere-Geosphere Interface* and *Thermochronological Methods: from Palaeotemperature Constraints to Landscape Evolution Models* (heaven help those of us who were taught not to invent new words). Only a very foolish person would suppose Huxley incapable of coping with these unusual, linguistically mysterious titles and the summary paragraphs that supported them, alien though the jargon would have been from his point of view. He would have retired to his study, inhaled a few cubic metres of tobacco smoke, written a brief note to *Nature* or *The Medical Times & Gazette,* and perhaps the next day remonstrated with the director of the Geological Society Publishing House (had it existed) over a dram of Glenlivet in the Athenaeum. Nonetheless, the substance, methodology and terminology of these two articles, and others like them, would have been unintelligible to him, particularly because the technology on which they are based would have been totally unfamiliar, which is simply a reminder that the idealistic, well-intentioned pioneer group that met in Buffalo should not be faulted for its members' innocence. They could not possibly anticipate how their own discipline would develop during the next a hundred and forty years and or that the professional, discipline-related language familiar to them would quickly become obsolete.

One of the talks Huxley gave while in Buffalo was later given the title of "Impressions of America". He coolly told his audience that though "he had heard of the degeneration of your stock from the English type", he had not himself observed degeneration. "I have seen no trace of the North American type, to which, it was said, you were reverting. You have among you the virtue which is most notable among savages, that of hospitality. I have visited your wigwams—and they are pretty good wigwams too. You entertain us with your best, and not only give us a good dinner, but are not quite happy unless we take the spoons and plates away with us".[44] The rest of his brief speech was devoted to remarks on the differences between English and American universities and to a report on what Professor Marsh had shown him at Yale. Of Marsh's collection he says: "There does not exist in Europe anything approaching it as regards extent, and the

[44] p. 76

geological time it covers, and the wonderful light it throws on the problem of evolution".

Interestingly, the reporter who recorded what Huxley said incorporated a brief description of him which ends as follows:

> In person Prof. Huxley is rather above the medium height; of large frame, but spare. He stoops slightly, as if habitually engaged in thought. His features are prominent and bear an expression of energy in repose. His hair and whiskers are iron grey. He speaks without manuscript or notes of any kind, and never prepares the phraseology of his addresses in advance of their delivery. His manner of speaking is quite simple and straightforward, with none of the gestures or arts of oratory. His delivery is slow and distinct, being the result of a hard and successful effort in the early part of his career to break off a previous habit of rapid speaking".[45]

Obviously the reporter must have had a chat with him afterwards.

The second talk Huxley gave in Buffalo, perhaps the second of several, was almost certainly for a general audience, not one just for scientists. With a title "From the Hut to the Pantheon: a Study of Evolution", it has been conserved in Scrapbook #2 in the Bolton-Stanwood papers in the Library of the American Antiquarian Society in Worcester, Massachusetts, that is to say, the manuscript of the talk as revised is conserved there. Once again it is interesting to have evidence of how Huxley customarily worked, as a spoken address delivered with only a few notes in hand and perhaps not fully thought out in advance was converted into something that was later published. Two types of foolscap paper are used in the manuscript now in the AAS; blue and cream. The statement "I hope that the American youth whom I address …" is on blue paper (p.4) while "when I was in Rome last year" is on cream paper (p.5). The water-mark of the blue paper is "Sawston 1881"; he could not have had it in 1876. We thus have another opportunity to see Huxley at work. He gives a roughly-hewn almost

[45] The adoption of a deliberate and slow manner of speaking was part of Huxley's attempt to be "popular" even if that meant speaking down to his audience. As often noted during his lifetime, he chose to talk as though most members of any audience would be totally ignorant.

impromptu talk in Buffalo in 1876 which he then revises for publication in London, at some point after 1881, in the magazine *The Youth's Companion.* Mario di Grigorio gives the date of publication as 1886. In the text as published there are references to Hawthorne's *House of the Green Gable* and *The Scarlet Letter.* Because these references are on blue paper, a case could be made for arguing they must have been introduced when Huxley was re-writing his talk in his den at home, but such an argument would not be conclusive, since he would almost certainly have said things in Buffalo that he did not write down. On the other hand, would he have been bold enough to talk about *The Scarlet Letter* to youngsters he had not previously met?

The use of the word "evolution" in the title is obviously of great interest. In Buffalo he used the word to mean an observable, step by step progression from a simple form to a more complex or sophisticated one. How can such a progression be achieved? The illustration he uses in the talk is architectural. If the idea of architectural beauty can be represented by a cylinder and a dome, then it can be said that a mud hut is the precursor of the dome of the Pantheon; they are not all that different but one is more highly developed than the other. In the last sentence of the talk as published he says: "Thus, as the Pantheon lay potentially in the mud hut of the Stone Age, so it may truly be said that the germ of the Piazza of St. Mark, with its glorious duomo, was contained in the prehistoric pile dwellings". This is a sentence that was probably not written in Buffalo since it was only later that he went to Italy, though of course it was not necessary to go to Italy to have the idea. Whenever it was written, however, it seems to confirm, though only on a simplistic level in a talk to young people, that Huxley could still only imagine evolution in very basic structural terms and had not resolved problems of the physical and biological adaptation to environmental circumstance. The use of the word "germ" seems cavalier until one remembers the history of the word and the science lying behind it. Even if the contrast between prehistoric mud hut and Renaissance architectural masterpiece is accepted simply as an explanatory metaphor, it is perhaps not a metaphor that would help a young audience or reader appreciate the significance of "deep time", the time, they would have to understand, needed for evolution to work. That said, it was of course the case that Huxley did well to participate, as his hosts desired in, as it were, the periphery or fringe of the conference. Indeed, not just participate, but actually make a positive contribution more or less on the spur of the moment. Then, back home, he was commendably frugal with the work he had done and the notes he brought back across the Atlantic, whereas another sort of person, though equally earnest, might have thought of the Buffalo performance as simply water under the bridge.

From the old Clifton Hotel where Huxley and his wife stayed for part of the time, they were taken on a number of trips, including at least one to Niagara Falls. This famous hotel, the first of several on the same site, was on the Canadian side of the river and of course was not then surrounded by the many facilities, amusement parks, stores, travel agencies, motels and colourful crowds that now greet thousands of tourists every year. Huxley could not fly over the falls in a helicopter as is one of the preferred ways of seeing them nowadays but he did venture underneath them,—just as a tourist, not a scientist. Where he stayed when he moved across the river to Buffalo is not known.

The Family Visit in Nashville, Tennessee

Having crossed the Atlantic, he was keen to see his elder sister, Lizzie who had married many years earlier and had been living in the South—in Alabama—since her marriage. Ostensibly to save Huxley from the even greater heat further south, she travelled north to Nashville where both her daughters were living with their own families. Brother and sister had little in common except their good-will and shared desire to make the most of the visit. But what could be done for such a visitor? He was known to be irreligious, therefore potentially embarrassing in such an overtly god-fearing part of the country. Always prepared to write about matters of which he had no personal experience, he had written about the Negro but had no knowledge of multi-culturalism. As far as anyone could tell, he lacked an ear for any type of music. Because this was essentially a family visit, thus remaining private in the very matters one would like to know about (Were there prayers at meals and at bedtime? Were his nieces and their families devoutly, seriously religious? Was he introduced to friends of the family who took for granted ideas and beliefs he was spending his life rebutting?), not much is really known about how he got on with his sister and her family, except that he was taken to see Nashville's two universities, Fisk and Vanderbilt, the first already well-established as an institution destined to become more and more distinguished within the black community, and the second, relatively new, but with high standards and high ambitions. Huxley was induced to give a talk in the 700 seat Masonic Hall, where he gently took his audience through the history of geology, with references to the local landscape and to Niagara Falls, which, because, or partly because his sister's son-in-law was Editor of the local newspaper, was favourably reported. By all accounts, the talk was a friendly, low-key, off the cuff, unobjectionable performance. But a sterner challenge awaited Huxley in Baltimore.

Before we follow him from Nashville to Maryland, it may be worth restating how this book is designed. It is about the professionally motivated visit he made to America within a fairly narrow compass of time (1870-1880) with a focus on 1876. It is not a family history, though questions are bound to arise that might make the reader want one. Only an entirely different book could satisfy a reader who wished to know why Huxley so rarely mentions his wife even when they are travelling together, why he keeps to himself whatever feelings he had about his sister and her family, and why he stays well clear of the questions of heredity that arise in any discussion of evolution but which he seems to avoid when thinking about the genetics of his parents' family as well as of his own.

Huxley's Address at the Opening of the Johns Hopkins University

Huxley chose to make higher education the subject of the address he had been invited to give at the opening ceremonies at the Johns Hopkins University, rather than, say, "the present state of scientific research" or "science in the international context". He had been in touch with Dr Gilman, the University's first President, before leaving Britain; indeed one of the few surviving letters related to this trip indicates that it was he who took the initiative, for on 23 April he had written to the President a rather disingenuous letter part of which reads: "I have just made up my mind to pay America a flying visit this autumn and hope to be able to spend a day in Baltimore, which has the reputation here of being one of the most charming cities in the States. I hope I shall be fortunate enough to meet you and see what the new university is doing".[46] There must certainly have been further correspondence between the two men before he was formally invited, so it can be assumed that the subject of the at that point unwritten address was mutually acceptable, even though he obviously cannot have known very much about American educational practices. Huxley thought he could give them the advice they needed. Dr Gilman later emphasised that he and he only had been responsible for the invitation.

Though completely ignorant of how things were done in North America, Huxley had for many years deliberately and conscientiously involved himself in British education, at least in England. Largely self-educated himself, therefore understanding at first hand the social obstacles to personal advancement that still existed in his own country, he had already, with many others, devoted time and energy to increasing educational opportunities for greater numbers of people, while at the same time endeavouring to decrease the dominance of the ancient, long-established schools and universities. His willingness to be involved in a practical way was illustrated by his briefly serving on a London School board

[46] ALS: Huxley to Gilman, 23 April 1876, Milton S. Eisenhower Library, the Johns Hopkins University.

until ill-health and the over-commitment forced him to resign. Together with a number of like-minded, public-spirited colleagues he gave lectures for "working men" in London, which were subsequently printed and sold for four pennies a copy. The books he produced at this time repeatedly demonstrated an awareness of how difficult it was, though how necessary, to make unfamiliar, seemingly complex subjects available and intelligible to the uneducated, or only partially educated. It was a continuing personal crusade to which he remained fully committed throughout his whole life. It was also the background to his invitation to the Johns Hopkins University. Perceiving the need for a wider, less exclusive educational system in Britain, he was prepared to devote his best energies to creating it, or at least to creating the conditions in which popular education could be achieved. These efforts coincided both with an increased awareness in the country as a whole of the drastic all-encompassing reforms that the country needed and with measures then being taken in Parliament to make the educational system less exclusive and more effective. During the years leading up to his American visit Huxley had been steadfastly on the side of radical reform—that is, not political reform, but reform by way of significant changes to the educational system so that those who wanted to improve their lives might have the opportunity to do so. Out with the old, in with the new. He saw science, in the most general sense, as uncontaminated by class power, social privilege or inherited wealth, believing more in the equalities that could be achieved by work while also perceiving that an increasingly powerful, inventive industrial nation needed a modern educational system that would support it. The decision to make education the subject of his address at Johns Hopkins was a deliberate not a casual one.

The speech Huxley delivered in Baltimore on 12 September 1876 was, however, very far from being the keynote address it should have been. He chose to pontificate in very general terms about both primary school education and university education without having had adequate experience of either. His own education had been fragmentary; he had never attended any university; he knew nothing about either historical or recent educational developments in the United States. In the speech as delivered, he said:

> Now I have a very clear conception as to what elementary education ought to be; what it may be when properly organised, and what I think it will be before many years have passed over our heads in England and America. Such education should enable an average boy of fifteen or sixteen to read and write his own language with ease and accuracy, and with a sense of excellence derived from a study of our classic authors; to have a general acquaintance with the history of his own country and with the great laws of social existence; to have acquired the rudiments of the physical and psychological sciences, and a fair knowledge of arithmetic and geometry. He should have obtained an acquaintance with logic rather by example than precept, while the acquirement of the elements of music and drawing should have been a pleasure rather than work.[47]

Whether true or not, this is not exactly riveting stuff. He lacked the close, day to day experience of any actual educational system that might have helped him determine whether these very general, idealistic but dogmatic remarks could ever have a practical application. Inappropriately his model was British; yet even in the British context, he overlooked the grammar schools that had existed since the fifteenth century in many parts of the country, the Church of England schools that continued to provide elementary education for a good part of the population, the many independently endowed schools, and of course the Public Schools where the curriculum he so disliked had been in place for several decades. He literally did not know what he was talking about. It will be remembered, though, that when he says "the primary school and the university are the alpha and omega of education", he is referring to a system in which the school leaving age in Britain was fourteen, while admission to a university was possible at a much younger age than it is nowadays. His single sentence on secondary schools is extraordinarily nonchalant. Huxley's prejudice about the universities of Oxford and Cambridge

---

[47] See *American Addresses* (New York: D. Appleton & Company, 1877), pp. 101-2

was so great he was unlikely to mention them, a prejudice, if that is the right word, stemming at least in part from his strong distaste for the connections that existed in those days between individual colleges and Church of England "livings" with which Fellows were often associated, but of course there were other universities he might have mentioned had his subject been legitimately British, not least King's College and University College, already flourishing, though still to be amalgamated within the University of London, and likewise the ancient Scottish universities of which he had some slight experience, which on this occasion he chose to overlook or forget.

Characteristically he expressed his opinions in the clear, uncomplicated, uncompromising, slow and dogmatic manner that had served him well on so many occasions, but had he forgotten that he was addressing an audience of two thousand well-informed, intelligent and strongly supportive guests at the opening of a new American university? Had he bothered to think out what he believed a modern university should be like? The answer is perfectly straightforward. He had not bothered. Or better, it could be said he *seemed* not to have bothered, or simply did not have the energy to do a good job, given his own physical condition and the rigours of their whirlwind tour.

It is important to have in mind that the text that was published in the book called *American Addresses* is by no means a record of what he actually said in Baltimore on 12 September 1876. The quotations and references above are taken from proof-sheets preserved in the Department of Special Collections in the Milton S Eisenhower Library at Johns Hopkins, proof-sheets that were probably prepared for newspaper release, possibly very soon after his departure. In as far as Huxley seriously misjudged the opportunity he had been offered, there is little point in engaging in a detailed analysis of these proof-sheets, vastly interesting though they are, but two points have to be mentioned because they seem to reveal something of Huxley's attitude of mind. The first concerns his appreciation of the likely rapid growth of the American population and its possible consequences. On this he writes the following rather grim, very revealing paragraph:

> You and your descendents will have to decide whether this great mass [of people] will hold together under the forms of a republic, and the despotic reality of universal suffrage; whether state rights will hold out against centralization without separation; whether centralization will hold out without actual or disguised monarchy; whether shifting corruption is better than a permanent bureaucracy, and as population thickens, and the pressure of want is felt, the gaunt spectre of pauperism will stalk among you, and communism and socialism will claim to be heard.[48]

Taking your host by the throat in this way was not exactly diplomatic. A whole thesis could probably be written on the basis of this paragraph alone, both in a consideration of phrases like "the despotic reality of universal suffrage", which seem to place him unambiguously far to the right of centre, providing an entry to detailed discussion of his right-wing political opinions and their implications, and also in a consideration of how he got on or did not get on with other human beings, socially and in the work place. Furthermore, when he opposes "shifting corruption" to a "permanent bureaucracy", his anticipation of a problem he thinks his hosts will have to face seems particularly indelicate. The permanent bureaucracy did not yet exist; did he think he had observed "shifting corruption" as he had travelled about? Because this paragraph was retained in *American Addresses,* it can presumably be taken as a clear statement of what he thought. Here then is another case of Huxley's retention of ideas and attitudes acquired early in life, in this instance most likely from John Stuart Mill's *On Liberty* which celebrates the individual's right to conduct his affairs in whatever manner he chooses, provided no other individual is harmed by the choices he makes.

The second matter that cannot really be avoided is the sexism of the whole address, which reads as though he believed not just that women could not and should not be educated, but even as though women did not exist as worth the

---

[48] ibid. p. 126

slightest consideration. This will nowadays grate on most readers' minds. One knows that up to the Second World War, and even beyond, there were lecturers at Oxford and Cambridge who found it amusing to begin their lectures with "Gentlemen" or "Good morning, gentlemen" even though there were female students in the lecture room clearly in view, but amusing, as a matter of fact, it was not. In Huxley's case there was not the slightest element of irony; he had simply never grasped the idea that women should enjoy, or might legitimately wish for the same educational opportunities as men and, when he gave his speech, did not know whether or not women would be admitted as equals for study at the Johns Hopkins University. On the home front, his own daughters, Jessie (aged 20 in 1876), Marion (17), Rachel (15), Henrietta (13) and Ethel had had to accept their place in nature—that is their place in their father's home as he wanted it to be run—their only options being the same as other women in Victorian Britain, elopement or marriage.

By far the most important discrepancy between the lecture as delivered and the book as published concerns what Huxley chose to say about medical education. The large, immensely generous bequest that constituted the initial funding of John Hopkins had specified that the money should be divided into two equal parts, one half for the university, the other half for the medical school, so it was strange and impolitic for the speaker to devote hardly more than a hundred and fifty words to medical education. Here is what he said:

> Lastly comes the question of how the university may co-operate in advancing medical education. A medical school is strictly a technical school—a school in which a practical profession is taught—while a university ought to be a place in which knowledge is obtained without direct reference to professional purposes. It is clear, therefore, that a university and its antecedent, the school, may best co-operate with the medical school by making due provision for the study of those branches of knowledge which lie at the foundation of medicine.
>
> At present, young men come to the medical schools without a conception of even the elements of physical science; they learn, for the first time, that there are such sciences as physics, chemistry and

> physiology, and are introduced to anatomy as a new thing. It may be safely said that with a large proportion of medical students much of the first session is wasted in learning how to learn—in familiarising themselves with utterly strange conceptions, and in awakening their dormant and wholly untrained powers of observation and manipulation.[49]

This was the gist of what he had to say, though the next two paragraphs of his speech, as reported, were devoted to the expansion and exploration of the one basic idea.

In the revised text as published, which is entitled "Address on University Education" without any mention of the Johns Hopkins University, there are, within the total of twenty eight book pages, eight entirely new pages on medical education introduced to precede those mentioned above, so that there is now a balance between university and medical school education. He had learnt his lesson, either from being criticized by people who had heard his address or from pondering what he had done as he returned home. From London he told Gilman: "I found it desirable to publish it in England, so I corrected it for *Nature*".[50] The publication of the text in North America had become problematic. Youmans wanted it for Appleton; Huxley seems not to have wanted to have anything more to do with the Johns Hopkins University. Eventually, when the drastically revised text was published by Appleton, there was no mention in it of the university where it had been delivered. In his letter to Gilman, Huxley had said: "I was rather shaky for two or three weeks after my return, but I am quite right again now, which is just as well, as I am head over heels in work".[51]

---

[49] ibid. pp. 116-7

[50] All the letters from Huxley to Gilman quoted in this book are in the Eisenhauer Library at the Johns Hopkins University. Huxley was a British imperialist so seems not to have appreciated that it would have been both courteous and professionally correct to allow his address to be published in Baltimore first.

[51] ALS: 11 November 1876. "I found that it was desirable to publish it in England so I corrected it for *Nature* and have now been over it again". Whereas someone else might have given his American hosts first refusal, Huxley invariably preferred to be first published in London, especially in this case because of the revisions people in Baltimore did not yet know about.

In this revision he retained the idea that a university should be a "liberal" institution entirely without courses aimed at professional qualifications, completely unlike a medical school in this respect where the whole point was to achieve a high level of professional competence. The word "liberal" as Huxley used it is a bit of a problem because many people use it differently nowadays, some even thinking of it as a term of abuse. It referred, in those days, to the type of education that was not aimed at any particular career but had more to do with self-fulfilment through learning, being in tune with the dominant contemporary culture in literature, music and the arts, and seeing civilized societies as driven, not by political institutions, but by the intellectual eminence of individuals regarded as the best safeguard against mob rule and social anarchy.

Huxley was well acquainted with Matthew Arnold, was a member of the same London club and would meet him for tea from time to time and very possibly at other times as well. Naturally he was familiar with both Arnold's *Schools and Universities* (1868) and his *Culture and Anarchy* (1869). In a typically nineteenth-century English fashion the anarchy they both feared was the transfer of power from the highly educated minority (the meritocracy but definitely not the aristocracy) to the as yet uneducated, politically restless majority, even though they both believed in the extension of popular education as the best means of keeping the masses under control. Matthew Arnold believed in the Greek and Latin classics as the means of achieving cultural stability; Huxley believed in science as having a similar social importance; they both feared and held away from the idea of democracy. When he said, towards the very end of his speech, that "universities ought to be, and may be, the fortresses of the higher life of the nation", he meant by the word "higher" culturally superior not politically more advanced than societies in other parts of the world, while with the word "fortress" he implied something exclusive and elitist, not inclusive and democratic. And when he said, in the same paragraph, "the one condition of your success is the moral worth and intellectual clearness of the individual citizen", he was confirming his position in the intellectual tradition of Bacon, Hobbes, Locke and Hume where it is not assumed on logical grounds that individuals will be capable of working together as equals except when endangered or forced to do so. Generally speaking, because Huxley was taking an entirely intellectual or perhaps rather shallow theoretical approach to his subject, as opposed to the approach that might be taken by a social scientist basing his or her observations of social conditions as scrupulously collected, measurable evidence, his idea of "moral worth", unsupported in the text, seems some sort of pie in the sky concept unrelated to the actuality of medical education in any part of the world. This notion turned out to be Huxley's Achilles heel. There was no reason to suppose human beings would behave well rather than badly; he was simply asserting a middle-class opinion that it would be better for families like his own if people behaved well,—that is, avoided criminality, accepted their place in society and respected other people's property. Besides, it has to be doubted that the Johns Hopkins audience wished to be lectured on "moral worth". That was something they already had.

The reaction to Huxley's speech in Baltimore was severely negative. For much of the audience it was a disappointment, lacking energy, relevance and any sense of celebration. For the local clergy it was particularly tasteless. One of them, who had no idea, apparently, that a university might be a secular institution, complained that the address should have started and ended with a prayer, and Daniel Gilman had to appease him as best he could. S.J. Prince, in a letter to the Rev. John Layburn, said: "Our Colleges are almost all born of the Christian religion, and it is sad to think that the latest born would have its first word said by that blatant infidel Huxley".[52] The idea of having a public institution without religious underpinning was difficult to grasp. And Thomas Clifford had written a similarly virulent letter to John Tyndall a year earlier, proclaiming that neither

[52] ALS 13 October 1870, Royal Institution #2941

Tyndall nor Huxley deserved more attention than "any other arrogant pretentious workmen".[53] Possibly the lecturer was almost as dissatisfied with what he had done as his critics. He had failed to make the best of a marvellous opportunity, had perhaps chosen an inappropriate subject and would have been well advised to spend time with Dr. Gilman before his performance. Throughout his American tour there was frequently a whiff of condescension as though he thought anyone not British was necessarily inferior. He seems to have left Baltimore rapidly and unceremoniously.

If the Johns Hopkins address is compared to other Huxley pieces on education such as "A Liberal Education and Where to Find It" or "Universities, Actual and Ideal", it looks as though what let Huxley down in Baltimore was a significant change in rhetorical stance, since, besides talking too much about Britain and not nearly enough about America, he chose not to utilize his normally preferred method of putting the negative first in a sort of Aristotelian manner and only then, if at all, presenting his positive ideas on whatever the subject happened to be. But on this occasion, Huxley's preparations had not been adequate. Youmans understood this when he read the report of the speech in the *Tribune.*

"A Liberal Education and Where to Find it" is presumably a reliable guide to what Huxley might have said in Baltimore had he been on form. After a very strained, affected chess metaphor which does not really add much to his argument, Huxley says in that essay:

> Well, what I mean by education is learning the rules of this mighty game. In other words, education is the instruction of the intellect in the laws of Nature, under which name I include not merely things and their forces, but men and their ways; and the fashioning of the affections and of the will into an earnest and loving desire to move in harmony with these laws. For me, education means neither more nor less than this. Anything which professes to call itself education must be tried by this standard, and if it

[53] Ibid

> fails to stand the test, I will not call it education, whatever may be the force of authority, or of numbers, upon the other side.[54]

This paragraph is full of interest. In the first place, Huxley is back on home ground. He can pit himself, rhetorically, against the "other side". For Huxley there must always be an opposition and the opposition will always be wrong, even when its opinions are not specified. Secondly, when he uses the word "nature" and gives it a capital letter, he is to a certain extent referring back to Wordsworth and Wordsworth's contemporaries, where "nature" means the external world and a human being's experience of it as environment, insights so brilliantly expressed in *The Prelude*; he also means the scientific laws to which human beings should, he thinks, submit, as though submission to circumstance, if not mandatory, is the best underpinning of a sensible life. This opinion, if it stood by itself, would represent an unattractive form of determinism. The phrase "men and their ways" cannot be expected to carry such a load. A criminal or a murderer might well believe he is successfully adapting to circumstance. Still, in this paragraph we have quintessential Huxley, quite unlike the Baltimore address where he so badly failed by his own high standards.

"In short", Huxley continues, "all artificial education ought to be an anticipation of natural education. And a liberal education is an artificial education which has not only prepared a man to escape the great evils of disobedience to natural laws, but has trained him to appreciate and seize upon the rewards, which Nature scatters with as free a hand as her penalties".[55] Sentences like these seem to demonstrate how close Huxley remained to Milton, an affinity which might have appeased the Johns Hopkins audience had he explained it and developed its implications. Active here is the idea of a necessary submission to fate, where fate is not the will of the gods as in Greek tragedy, or Christianity, or any other

---

[54] "A Liberal Education and Where to Find It" in *Science and Education* (London: Macmillan, 1899), p. 83

[55] ibid. pp. 85-6

religion, but science as the structural reality of an individual's life on earth to which it would be foolish not to submit.

Then comes one of the more justifiably famous passages of this particular essayist:

> That man, I think, has had a liberal education who has been so trained in youth that his body is the ready servant of his will, and does with ease and pleasure all the work that, as a mechanism, it is capable of; whose intellect is a clear, cold, logic engine, with all its parts of equal strength, and in smooth working order, ready, like a steam-engine, to be turned to any kind of work; and spin the gossamers as well as forge the anchors of the mind; whose mind is stored with a knowledge of the great and fundamental truths of Nature and of the laws of her operations; one who, no stunted ascetic, is full of life and fire, but whose passions are trained to come to heel by a vigorous will, the servant of a tender conscience; who has learned to love all beauty, whether of Nature or of art, to hate all vileness, and to respect others as himself.[56]

The Johns Hopkins audience might have liked part of this, though obviously not the whole. The idea that a person should control his or her instincts or feelings in order to "hate all vileness", while behaving towards his neighbours as they should behave towards him, is actually a moralistic, almost religious notion, which might have found favour with the Baltimore audience, but obviously has nothing to do with science. Huxley was grafting morality onto his atheism. Here, as elsewhere, he is quite unable to explain why someone educated in a non-religious manner that has an awareness of modern science at its centre should not be a criminal, a thief, or an adulterer. More than a few people with science degrees have in fact been criminals, thieves and adulterers. It was because of the illogicality of this position that he subsequently preferred to be regarded, not as an atheist, but as an agnostic, a semantic switch that allowed him to continue to pontificate about good and bad behaviour, as though understanding the "laws" of nature did not have ethical implications and was not at variance with the Christian creed. He was running with the fox while hunting with the hounds. To be fair to him, in the

[56] ibid. p. 86

speech as delivered, he did in fact talk about the relationship of morality and science, but unfortunately not in a very convincing or well-argued manner. The Johns Hopkins audience was not unintelligent. It deserved a more coherent, less didactic approach.

In the passage reproduced above, the image of the human being as a machine has a certain historical charm, in as far as the organic insights and metaphors of the European Romantic period had given way, by 1876, to the mechanistic images and concepts regarded as valid or unexceptional by many mid-Victorians, though it seems reasonable to observe in passing that the view of the human being as a steam-engine in good working order would seem to be invalidated by Huxley himself, who was so frequently ill (seized up?), and not reliably in good working order (often needing the medical or psychiatric attention that he sometimes received and sometimes did not). Still, machine metaphors certainly belonged to the period, as will be discussed in the next chapter. For the moment, the point is to express surprise that someone who held such strong convictions about education, as expounded not just in "A Liberal Education and where to Find It", but in many other articles and notes written roughly at the same point in his career, had such difficulty in putting together a powerful, appropriate talk for the Johns Hopkins audience. Although it has been suggested that he put himself at a disadvantage by having to manage without his notes,[57] this cannot be the whole of the story, and indeed may not be accurate, since he routinely lectured without notes of any kind. Physical exhaustion was probably part of the cause of failure, relatively speaking. He had been travelling for several weeks. There had to be limit to what he was capable of doing. He had reached that limit.

---

[57] Adrian Desmond in *Huxley: Evolution's High Priest*, p. 96, says that Huxley had intended to have in hand the script he had dictated on the previous day to an Associated Press stenographer, suggesting that because the copy had not reached him in time he was unable to recollect the text as he knew it was going to be published. This anecdote has to be doubted simply because Huxley very rarely, if ever, lectured with his notes in front of him. Still, the episode is certainly open to a variety of interpretations.

Washington and Philadelphia

From Baltimore the Huxleys travelled by way of Washington to Philadelphia where they stayed in the Continental Hotel for one night. From the hotel Huxley sent a short note to Spencer Baird (1823-1887) whom had met in Buffalo where he, Professor Baird, had suggested they might go together to the International Exhibition of Arts, Manufactures, and Products of the Soil and Mine. "I shall be very glad to avail myself of your kind offer. My wife and I will be at your office at ten o'clock tomorrow morning as we have to leave for New York in the afternoon".[58]

This was the Centennial International Exhibition that had opened on 10 May just a little too late for the actual centenary of the battles of Lexington and Concord. Opened by President Ulysses Grant, the exhibition was designed to commemorate the Declaration of Independence. It is not certain how many of the 30,000 displays the Huxleys could have seen. Situated on 450 acres of parkland bordering the Schuylkill River, the exhibition was vast, both as designed to show American manufacturing at its best and to display modern products from more than twenty other countries. The main exhibition hall, designed by Henry Pettit and Joseph Wilson, had a floor space of twenty acres. Among the thousands of exhibits were bicycles (one of the very earliest penny farthings), a telephone (Alexander Bell had been there), a display of Remington typewriters, and an electric dynamo, as well as samples of other consumer novelties such as root beer and Heinz ketchup. Designed to show off contemporary American inventiveness and technological genius, the huge hall also contained an impressive collection of all types of modern machinery—for farms, mines, modern cities and all forms of transport system. Which items and which parts of this huge exhibition attracted the Huxleys' attention has not been determined. Seeing the typewriters, if he did, failed to influence Huxley's only too well established work habits which he chose to retain for the next thirty years. It is quite impossible to imagine him typing and

[58] ALS: 14 September 1876 (wrongly catalogued as 1875?), Smithsonian

re-typing his own work. He was always much too busy. It also has to be doubted that he ever tasted tomato ketchup. As they could only afford to be there for a few hours, they probably did not see either the Women's Pavilion or the splendid Horticultural Hall.

Huxley and Baird remained in touch with each other for many years. They were both scientists with wide interests and serious, long-term commitments to the practical administration of science in relation to public life and government. In 1845 Baird had been appointed professor of natural history at Dickinson College, but had soon moved to an important administrative-cum-scientific post at the Smithsonian Institution, where he remained for the rest of his working life. From 1871 onwards he had also been in charge of the United States Fish Commission. His publications included *Catalog of North American Reptiles* (1853), *Birds, in a series of Reports of Explorations and Surveys for a Railway Route from the Mississippi River to the Pacific Ocean* (1858), *Mammals of North America: Descriptions based on Collections in the Smithsonian Institution* (1859) and *History of North American Birds* (the first of several volumes, being published in 1875). Though Baird was well ahead of Huxley in work-related travel, they had a great deal in common, both serving on fisheries commissions, both seeing the importance of institutions for the advancement of science, both writing and publishing a great deal and both being well-placed in London and Washington to play influential roles in the development of science policies. Their next meeting would be in Britain.

The Three Lectures in New York

The Huxleys spent the last five days of their American trip in New York City where Huxley delivered three lectures on 18, 20 and 22 September, which he would later publish as "Three Lectures on Evolution" but which the *New York Tribune* entitled "The Direct Evidences of Evolution". In the special issue put out by that newspaper there is an interesting description of the occasion.

> Within, every seat seemed to be taken before the lecture began, the few vacancies were filled during the first ten minutes afterward, and 'wall-flowers' were packed standing behind the seats. The hall was full of familiar faces, of men eminent in the learned professions; of New York's best society. Punctual to the very minute Prof. Huxley came forward upon the platform, and was of course greeted with abundant applause. He laid a copy of Milton's *Paradise Lost* upon the reading desk; nothing else, neither manuscript or notes. He leaned forward slightly over the desk and began speaking in measured words and with a low tone of voice. Except sometimes to gasp the desk with both hands and lean over it more intently, he did not vary his position or make use of gestures during the lecture.[59]

Though the newspaper advertisements in advance of these lectures have not been tracked down, they were obviously effective for there to have been such a crowd. His reputation had again preceded him. A large number of people wanted to hear the great man and to be present at an important occasion. Precisely what they heard is difficult to determine. As usual there is a discrepancy between the lecture as delivered and the text as published, that is, if the reporter was being accurate who said that the first lecture took an hour and a quarter to be delivered, and if Huxley had reverted to his preferred practices.

As said earlier, his preferred mode of operation in the lecture room was not to have a written text from which he read or notes to which he could refer if need be, because he would then have needed his reading glasses, but to have prepared with sufficient care and thoroughness for the rhetoric to carry him through, rhetoric here meaning a coherent structure of ideas, arguments and illustrations which would act as mnemonic for this highly intelligent man who happened to have an excellent memory. In the published version of this and other talks, the rhetorical structure provides a coherence which is helpful to the reader even though it severely diminishes the feeling of spontaneity. From the point of view of an audience, the impression was of a serious, thoughtful, well organized

---

[59] *The New York Tribune* Extra Issue

and knowledgeable scientist sincerely attempting to convey his thoughts to a general audience in an intelligible and helpful manner.

In these three lectures Huxley is in effect walking an intellectual tightrope. He wishes to approach the subject of evolution without giving the impression that he is merely attacking religion, so he does what he is good at, addressing the negative first in an almost Aristotelian way, before coming to the positive side of the argument in his last lecture. This is like going to someone's house for a chicken dinner. You know there is going to be chicken but the host says: "I've got a lovely bird to cook but I wonder what's the best way". Then you hear about the various ways in which a chicken might be cooked, as well as the ways which over the years have been rejected, much of which you already know. "I'm not too keen about that Moroccan dish we had at Sally's, and as for the way they do it at Gilbert's restaurant downtown… well!" During all this time, you know a chicken dish is already in the oven—you can smell it, just as someone reading Huxley's *Lectures on Evolution* knows at the outset he will be proposing some version of modern (for him) evolutionary theory. That, after all, is why he had an audience. But the rhetoric is so unlike the conversations you might hear on the same subject, in a bar, or a locker room, or a downtown coffee shop that it has a certain old-world charm, as the evident evasiveness, or the strategy of pretend evasiveness, has charm. When all is said and done, however, the lectures were intended to explain why evolutionary theory had to be accepted as overriding all other views of human existence and why religious and metaphysical ideas and practices, however cherished, must be abandoned because demonstrably invalid..

At the beginning of the first lecture he says: "it has ceased to be conceivable that chance has any place in the universe". It is extravagant statements of this kind that locate him firmly in the third quarter of the nineteenth century, and at the same time hint at some kind of intellectual limitation. Huxley was not thinking clearly when he made this remark. What he meant was that all types of hocus-pocus, as he regarded it—mysticism, prayer, superstition, belief in a god, and so on—had been replaced by science where either something had been defined absolutely and could only be "known" in that context or, relative to this "something" and it only, the limits of knowledge had been clearly defined. There was simply no place in Huxley's way of life for any type of knowing different from the one he had espoused. Chance cannot be so readily brushed aside, however, either in a day by day sense or in a scientific sense.

Of the three lectures, the first was about the impact of geology on belief; the second was about the understanding of palaeontology as having a bearing on how the world might best be viewed as opposed to any view based on the Christian Bible; and the third was about the idea of evolution as founded on the discovery of sequences of anatomical variations over long periods of time. He starts with geology in order to insist both that it can be shown there was life on earth a long time before the modern era began, and therefore before Christ, and also that there was a time when the earth was without life of any kind. So much for religious creation myths.

In the second lecture, he resorts to talking about Milton's *Paradise Lost* instead of talking directly about the implausibility of the Christian story, and of the Book of Genesis in particular, because, in the Milton passage he quoted, he can reasonably suggest that what Milton wrote is patently untrue, though he covers himself by saying "for my part I have no prejudice one way or another", which was a disclaimer forced on him by the reception he had received in Baltimore. Chapter Two is disarmingly entitled "The Hypothesis of Evolution. The Neutral and the Favourable Evidence". In a very guarded set of statements, he cautiously gives credit to Darwin's ideas on biological variation and on what he calls the "influence of conditions" as against his (Huxley's) normal insistence on "persistent types", that is anatomical forms that remain unchanged for century after century.

Chapter Three is about the evolution of the horse as an example of anatomical variation and adaptation over time. Most of the information in this lecture is taken directly from what Othniel Marsh showed him at Yale, so much so that there is no need to recapitulate it here, since the detail has already been given earlier in the present chapter. This is an occasion when he *does* glancingly acknowledge the work of others, particularly in the following passage:

> Of late years there has been discovered in your Western Territories that marvellous accumulation of deposits, admirably adapted for the preservation of organic remains, to which I referred the other evening and which furnishes us with a consecutive series of records of the fauna of the older half of the Tertiary period, for which we have no parallel in Europe. They have yielded fossils in an excellent state of conservation and in unexampled number and variety. The researches of Leidy and others have shown that forms allied to the *Hipparion* and the *Anchitherium* are to be found among these remains. But it is only recently that the admirably conceived and most thoroughly and patiently worked-out investigations of Professor Marsh have given us a just idea of the vast fossil wealth, and of the scientific importance, of these deposits. I have had the advantage of glancing over the collections in the Yale Museum; and I can truly say that, so far as my knowledge extends, there is no collection from any one region and series of strata comparable, for extent, or for care with which the remains have been put together, or for their scientific importance, to the series of fossils which he has deposited there. This vast collection has yielded evidence bearing on the question of the pedigree of the horse of the most striking character.[60]

The Leidy mentioned in this paragraph is Joseph Leidy (1823-91), at one time a professor of anatomy at the University of Pennsylvania, who distinguished himself in several different fields of enquiry, one of them being palaeontology. It seems likely that Huxley would have known Leidy's *Extinct Fauna of Dakota and Nebraska* (1869), though perhaps not before his North American visit. Huxley refrains from mentioning Leidy's distinguished pupil, E.D. Cope, though probably not from ignorance. He would have known that the two palaeontologists, Cope and Marsh, had been rivals for many years.

---

[60] *American Addresses with a Lecture on the Study of Biology* (New York: D. Appleton and Company, 1877), pp. 85-86

Once home, Huxley carried his research a step further by corresponding with the French palaeontologist Lartet, nonchalantly mentioning, in *American Addresses* though without offering any explanation of his action: "I found myself corresponding with Lartet" (p. 83). If Huxley's correspondence occurred after his American travels, this must have been Louis Lartet (1840-1899) not his famous father, Edouard Lartet (1801-71) whose work, however, Huxley knew well, including *The Antiquity of Man in Western Europe* (1860) and *New Researches on the Coexistence of Man and the Great Fossil Mammifers Characteristic of the Last Geological Period* (1861). The phrase "I found myself" may, however, have been a fudge. When he went to see Marsh at Yale Huxley had not been completely certain what to expect; but when he had time to reflect on what he had seen and learnt, he would have remembered that what he knew of Lartet's work was distinctly relevant.

In what Huxley called the epilogue to his three lectures, he brought the series to an end in a characteristically gracious manner, which the last two paragraphs of his third lecture well illustrate.

> When I commenced this series of lectures, I did not think it necessary to preface them with a prologue, such as might be expected from a stranger or a foreigner; for during my brief stay in your country, I have found it hard to believe that a stranger could be possessed of so many friends, and almost harder that a stranger should express himself in your language in such a way as to be, to all appearance, so readily intelligible. So far as I can judge, that most intelligent, and, perhaps I may add, most singularly active and enterprising body, your press reporters, do not seem to have been deterred by my accent from giving the fullest account of everything I happen to have said.
>
> But the vessel in which I take my departure tomorrow morning is even now ready to slip her moorings; I awake from my delusion that I am other than a stranger and a foreigner. I am ready to go back to my place and country; but, before doing so, let me by way of epilogue, tender to you my most hearty thanks for the kind and cordial reception which you have accorded to me; and let me thank you still more for that which is the greatest compliment which can be afforded to any person in my position—

> the continuous and undisturbed attention which you have bestowed on the long argument which I have had the honour to lay before you.[61]

"Nothing could have been more interesting and enjoyable in every way than our tour of America", he would tell Moncure Conway when he got back to Britain, "and I shall never forget the thoughtful kindness and hospitality with which we were received".[62]

The newspaper reports of his lectures were certainly important because they contributed to people's knowing more about him in America than in Britain, which is not to say they were all favourable. The contrary was the case. It was probably the negative press reaction to his Baltimore talk that made him trim his sails in New York, where he camouflaged his unyieldingly strong and always negative opinions on Christianity by taking Milton's *Paradise Lost* as one of his key illustrations and where he protected himself with expressions like "he will abstain, as I do, from giving any opinion" (p.20) and "I shall deal with the matter as entirely a question of history" (p.29). He had learnt the hard way, particularly from sections of the press, that people, or some people at least, disliked being attacked for their religious beliefs by a person known to be an atheist. However clever his rhetoric, they found it impossible to believe the very basis of their daily existence could be called in question in this aggressive manner, besides which only a minority of them took the Book of Genesis literally as he had supposed, or had pretended to think they did. They did not think it was the sole basis of the Christian faith. They knew they were being attacked. No wonder that, before long, he would find it expedient to call himself an agnostic.

Postscript to Chapter Two

---

[61] ibid. pp. 95-96

[62] ALS Huxley to Moncure Conway. The tranche of letters from Huxley to Conway in the Butler Library at Columbia indicates that they remained friends for many years.

In a positive, good-hearted and genuine manner, Huxley had acquitted himself well during his five week tour, having done his best to respond with professional high competence to the various invitations he received and to the demands made upon his time, many of them unexpected. He had given lectures or talks in Buffalo, Nashville, Baltimore and New York; had travelled thousands of miles; had tested his limited travel wardrobe to the fullest extent, and had met many distinguished Americans some of whom became friends for life. He could hardly have done more. There were, of course, aspects of American life for which he had not been prepared. He had failed to anticipate the strength and depth of religious conviction among these descendents of the Pilgrim Fathers and he was taken off-guard when asked to enjoy an open friendliness and hospitality that did not have its equivalent in England, at least in the part of English society that Huxley knew. In so many ways the trip was nonetheless a great success, strengthening his international reputation, widening his horizons, and drawing attention to his many books that Appleton would continue to publish for the rest of his life.

Intellectually, however, he had backed himself into a corner, partly because of his egoism (he remained unwilling to be involved in shared enterprises) and partly because the intellectual position he was seen to be adopting was untenable.

It could be argued, possibly, that he did not mean, in his lectures and books, to give the impression he was the only person in the world who had such thoughts about religious faith and conviction, about the conflict between science and religion, and about the extension of time many millennia before the birth of Christ. Nonetheless, he quite often did give that impression. Because, when all is said and done, he was not a research scientist, he seems to have been unwilling, at least in North America, simply to talk about recent advances in geological and palaeontological enquiry. He wanted to be controversial. He also wanted to be his own man, unique, original, which meant that he would often choose to avoid or bypass what he must have known his audience would already know, despite all the discussion of evolution and its implications that had been in progress (or had been raging?) for the previous eighteen years and, indeed, longer. For example, many in his audiences would have known and thought about the geology passages in Tennyson's *In Memoriam*, but to have recollected them explicitly in his own work would have been out of character. Likewise, there was a good chance they would have known the popular science writings of Grant Allen, as published by Appleton, but Huxley may have thought his speeches would have been weakened by too much reference to other people's work, and perhaps he was right. Americans meanwhile were reading the same journals as he read himself and some of them, like him, attached importance to being up with the field—knowing what others knew. Huxley must have understood that other palaeontologists were thinking and writing about the very same problems he was thinking about himself. Would not more reference to them have been helpful, at least in his books? He had no need to expose himself as an out and out egoist. Ever since *The Origin of Species* had appeared, many, many people had applied their minds to the problem of whether it was possible to accept the findings of modern geology and palaeontology *and* remain a practising Christian. Surely he would have realised this.

Continuing to be a practising Christian did not necessarily mean subscribing to a literal interpretation of the Book of Genesis in the Christian Bible, indeed for many it meant believing in the important modifications, revisions and interpretations offered by the New Testament, so Huxley's remarks on the Christian account of the creation of the world in his New York lectures seemed gratuitously aggressive—a cheap shot, some might say, in the sense that a person sincerely trying to work out his or her position in the conflict between science and religion would have gone well beyond such a simple formulation as Huxley's. When Huxley visited America, the work of the several American Bible societies was in full flow. Some of these were dedicated to establishing literal interpretations of the Christian Bible, including the Book of Genesis, but many were much more sophisticated, including the British and Foreign Bible Society which confined its activities to the distribution of Bibles.

Because Huxley was not a Christian himself, it may be that he knew very little, or not enough, about the painful process experienced by many in the accommodation of new knowledge. Even if, like the present writer, a person is not religious at all, it cannot be seen as useful to over-simplify such an important conflict. A person confirmed into the Church of England, for example, has been examined in his knowledge of the Thirty Nine Articles, the Creed, various parts of the Prayer Book and other elements of Christian belief, so has been forced to think about matters generally considered to be important, and would very likely not be faulted for openly wondering how the Book of Genesis should be interpreted, in as far as very few thoughtful people would be bold enough (or silly enough?) to interpret it literally. This is not to say that geology and Christian belief can be convincingly reconciled. It is only to say that Huxley's version of the conflict, at least during his American tour, was unhelpfully simplistic and aggressive. He subsequently made the right decision when he took refuge behind the word agnostic. Being so straightforwardly identified as an atheist during his travels had had a negative effect on his reputation.

Huxley only rarely read a novel, did not see novels as having anything to do with faith, belief or intellectual conviction, and demonstrated little awareness of the fact—if such a word can be used in this context—that if a novelist wished to motivate a character in a manner an intelligent reader would find convincing, no good would come from attributing old-fashioned, traditional reasons for modern behaviour, such as the desire of a woman to marry a wealthy man or the desire of a man to possess a beautiful woman, since such reasons for a character's behaving in this way or that might serve adequately in popular fiction, but would be seen as naïve, out of date and tedious if the novel were being regarded as a serious literary form by the person making it. What does "serious" mean in this context? Among other things it would mean not writing as though theories of evolution had not affected the way in which a novelist might activate or motivate the principal characters in a work of fiction, and this might apply whether the characters were aware of their predicament or not. Roughly contemporary with Huxley's American tour were novels like Thomas Hardy's *Far From the Madding Crowd*, Henry James' *Roderick Hudson*, George Gissing's *Workers in the Dawn*, George Meredith's *Beauchamp's Career* and other strong, traditionally written novels like them, but a shift was in progress. Theories of evolution, the facts of extinction, the biological significance of heredity and a drastically reduced willingness to attribute meaning to the life of an individual together meant, not just that a change of stance or attitude had been forced on the serious novelist, but also that the psychological make-up had to be consistent with the new knowledge—either that, or the narrator would have to demonstrate an awareness of intellectual change even if the characters did not.

This awareness that the world a novelist's characters inhabited had altered in important ways shows up much more strongly in novels published only a short time later. In Britain any list of such novels would include Hardy's *Tess of the D'Urbervilles* and *Jude the Obscure*, Gissing's *New Grub Street,* and Meredith's *Diana of the Crossways* and *The Amazing Marriage.* In these books, either the narrator shows that what is motivating the characters is consciously post-Darwinian, or both the narrator and some of the characters show it. In this post-Darwinian world the characters behave differently; romance has been devalued; sex is a biological function; there is no reason to behave well in accordance with the apparently outdated ethical values of former generations; new patterns of behaviour have emerged that have altered the nature of the personal relationships as the novelist is able to describe them; women characters, particularly, think about themselves, their environment and the other human beings they encounter in a manner earlier novelists simply did not know about and could not have anticipated. Generally speaking, the novelists of this late nineteenth-century period, like the dramatists contemporary to them, absorbed the science as they absorbed much else seen to be relevant to their work, while only a small number of scientists absorbed the relevance of the novel, though, to appreciate the significance of post-Darwinian science, it was not necessary to come to terms with the sort of detail Huxley referred to in his New York lectures.

Of direct relevance here is George Gissing's novel called *Born in Exile* which was started before the novelist fully understood evolutionary theory, was then put aside and was only completed after he had read Théodule Ribot's book on heredity, which was published, incidentally, in the International Scientific Series. The novel is a study of the impact of geology on human behaviour. All the characters in this book are represented as understanding their very existence, their day to day functioning and their introspection, have been deeply affected and altered by concepts of evolution, to the extent they perceive they are not, after all, fully in control of their own lives, at least not in the ways they had been taught. What they were induced to believe was that evolutionary forces control human behaviour in a manner the individual can do nothing about. This particular novel expresses a philosophical determinism which Huxley, after America, saw he would have to reject but which the characters in *Born in Exile* believe they have no option but to accept. They know about the Bucknell Museum at Bordighera. Like Huxley, they have all read the works of Lyell, Darwin, William Buckland and Hugh Miller, as well as other less well-known books like Reusch's *Nature and the Bible: Creation in Relation to Natural Science* (1886) and Pfaff's *The Age and Origin of Man Geologically Considered* (1883), both of which are explicitly part of the fabric of *Born in Exile*, being mentioned as intellectual markers in the text. From the point of view of this particular novelist, accepting the intellectual implications of evolution meant seeing social relationships, and society itself, in a completely new light, since an individual's life could have no more meaning than that of an individual ant on an ant hill. After 1876, Huxley deliberately and thoughtfully tried to avoid the deterministic and nihilistic conclusions regarded as unavoidable by Gissing's characters in *Born in Exile*, but it probably fair to say that, while in America, he failed to see the difficulties he was making for himself.

On the other hand, he may have had a greater degree of self-knowledge than is suggested by his bold, but sometimes naive behaviour in America. Towards the end of his pocket-sized "Autobiography" he says:

> If I may speak of the objects I have had more or less definitely in view since I began the ascent of my hillock, they are briefly these: To promote the increase of natural knowledge and to forward the application of scientific methods of investigation to all the problems of life to the best of my ability, in the conviction that has grown with my growth and strengthened with my strength, that there is no alleviation of the sufferings of mankind except veracity of thought and of action, and the resolute facing of the world as it is when the garment of make-believe by which pious hands have hidden its uglier features is stripped off.
>
> It is with this intent that I have subordinated any reasonable, or unreasonable, ambition for scientific fame, which I may have permitted myself to entertain, to other ends; to the popularisation of science; to the development and organisation of scientific education; to the endless series of battles and skirmishes over evolution; and to the untiring opposition to that ecclesiastical spirit, which in England , as everywhere else, and to whatever denomination it may belong, is the deadly enemy of science.[63]

What, though, was the alternative to the ecclesiastic spirit? Was there an alternative? In "Evolution and Ethics" as printed in *Collected Essays*, Vol.9 (1894) Huxley can be seen still clinging to the notion that there is an alternative to what he calls "fanatical individualism" and "reasoned savagery" which, after such a long period of time, he still associates with his one-time close friend, Herbert Spencer. To discuss in any depth the differences between Spencer and Huxley, both of them denying the march of events in an unhistorical, non-political way, would require another book. All one can suggest at the moment is that Huxley's North American jaunt triggered a line of thought and intellectual accommodation that left him simply having to decide between implausible alternatives. Would he

---

[63] *Lectures and Essays* (London: Macmillan, 1910), pp. 10-11

be a moralist or a nihilist? His answer to this, in *Evolution and Ethics,* is extraordinary. He says:

> Social progress means a checking of the cosmic process at every step and the substitution of it for another, which may be called the ethical process; the end of which is not the survival of those who happen to be the fittest, in respect of the whole of the conditions which obtain, but of those who are ethically the best.[64]

Being seen as an atheist, a nihilist, a social rebel and an outsider, which for many years had been clearly implied in his belligerent, anti-religious attitudes, was the logical consequence, especially, of the opinions he had expressed during his American tour; he wanted to be an insider, a good citizen and a scientist whose work had social importance. But was this possible? He had cast the dice during his American travels. All he could do was to leave the table and not play that particular game ever again, or at least not in the same way.

[64] *Evolution and Ethics and Other Essays* (London: Macmillan, 1894), Collected Essays, Vol. IX, p. 81

T.H. Huxley, National Library of Medicine, Bethesda, Maryland

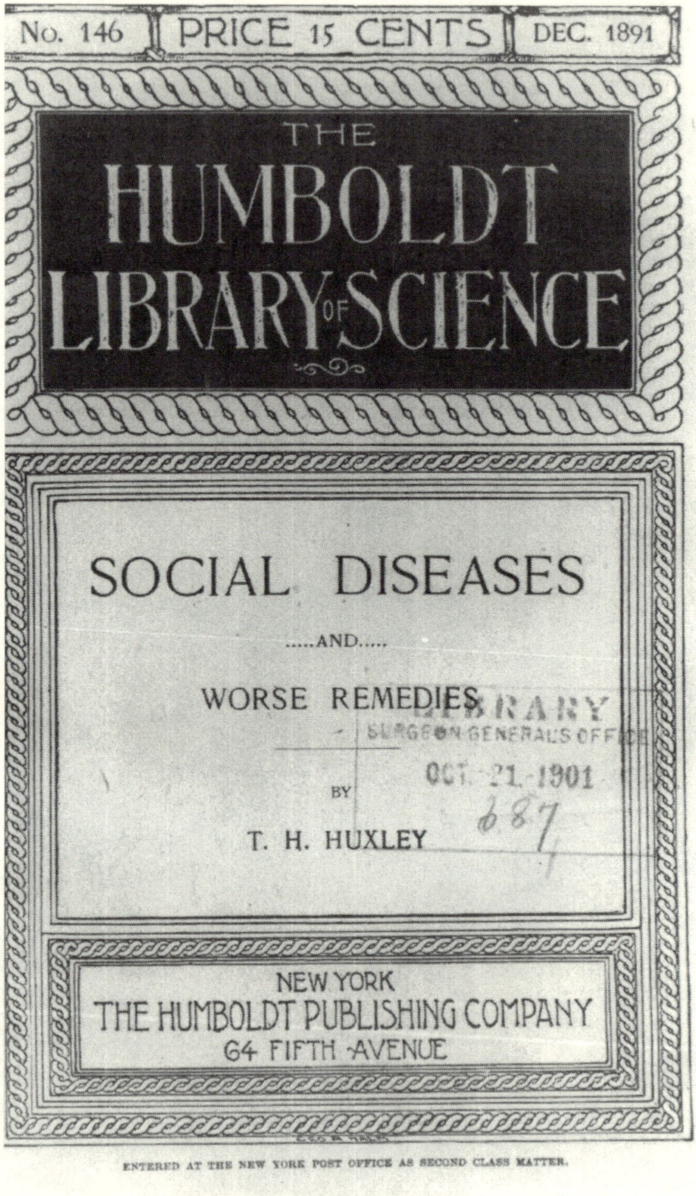

No. 146 PRICE 15 CENTS DEC. 1891

THE HUMBOLDT LIBRARY OF SCIENCE

SOCIAL DISEASES

.....AND.....

WORSE REMEDIES

BY

T. H. HUXLEY

NEW YORK
THE HUMBOLDT PUBLISHING COMPANY
64 FIFTH AVENUE

ENTERED AT THE NEW YORK POST OFFICE AS SECOND CLASS MATTER.

Cover of the Humboldt Library of Science Social Diseases and Worse Remedies
by T.H. Huxley

CHAPTER THREE

Of the two Huxleys who are the subject of this book one generally made at least a meagre attempt to acknowledge other people's work, while the other usually did not. This was a matter of style, method and personal choice, not of ignorance or carelessness. Huxley very rarely left a problematic subject open to a variety of interpretations, even when he knew a variety of opinions existed, preferring instead to suggest that, having demonstrated that all possibilities had been considered, he had come to the correct conclusion, that is to say the only conclusion logically possible. This preference for logic over induction provided the rhetorical structure of most of his talks. In other words, he preferred to adopt a position on any subject that was logically impregnable, even while admitting, quite often, that everything was not yet known, rather than present evidence based on research experience which could be variously interpreted.

Because of his clearly stated convictions about the importance of being logical, he seems to have believed, or to have talked and argued as though he believed, there would usually be one correct and several, or many, incorrect conclusions to the solving of any problem, where problem solving was taken as being a primary or *the* primary intellectual activity, so, because of this belief, tended to be intolerant of anyone who disagreed with him. To the extent that this was his habit, it certainly demonstrated an enviable strength of mind, since he had the knack of constructing apparently unassailable intellectual barricades, but, on the other hand, it perhaps also revealed a fairly serious psychological problem, one that must have gone back to his early education, where he was taught that thinking clearly was at least as important as being guided by whatever evidence could be gathered. His mind was not open to the possibility of there being several valid approaches to any subject. There were not thirteen ways of looking at a blackbird, as in the Wallace Stevens poem; there was one way. This had the advantage of his always appearing to be searching for the absolute truth (his reputation depended on this), but it had the disadvantage of holding him away from any other type of mental activity. He was very much better at monologue than dialogue, in any debate more wanted to come out on top than search for compromise and accommodation, and would prefer to lose a friend or make an enemy rather than admit that he might be wrong. This inclination always to be dominant, and the undoubted sincerity of the inclination, affected the way he wrote and the way he gave speeches, since there was rarely any room in them for the opinions, judgements, research findings or life-long, deeply considered convictions of other people. This was one of the two Huxleys mentioned above

An example may be useful. In the scientific work gathered together by Michael Foster and Ray Lancaster in *Scientific Memoirs of Thomas Henry Huxley*, Huxley was by no means negligent in noticing the published research of other scientists, though understandably the conventions he adopted were different from those nowadays current. To take just one instance of this (since what is being said applies to most of his other scientific publications), in the publication entitled "British Fossils Part 1. On the Anatomy and Affinities of the Genus Pterygotus", (*Monograph 1*, 1859, pp. 170-172), Agassiz, Webster, Murchison, Lyell and Buckland are mentioned by name in the preamble, where there is also a long quotation (pp. 170-172) from Agassiz's *Recherches sur les Poissons Fossiles,* but the references are not specific since Huxley is simply easing his way into his subject in a manner designed to show that a subject exists. The references, taken together, are no more than window dressing. What the individuals actually thought will not come into Huxley's treatment of the subject. In the next part of the Memoir (pp.173-75), he mentions Milne-Edwards, Dana, Bell, Baird, Schaffer and Darwin, as well as Page's *Advanced Text-Book of Geology*, while lastly, he says, "I stated the results of the same enquiries still more fully in my "Lectures on General Natural History", published in the "Medical Times and Gazette" for 1857". This may seem dry stuff, but the point being made is that Huxley, when referring to the work of other scientists, is not engaging in any kind of dialogue about how to interpret the evidence (there is no indication yet of what the evidence is), but is simply demonstrating that he is not unaware of the general field of enquiry to which he believes his monograph will be an authoritative contribution. This is like a man on a Mediterranean cruise ship saying, no, he had certainly not stayed in his cabin like a moody recluse during the two week excursion unaware of what was going on around him, but on the contrary had slept with seven different women, in *their* cabins, while aboard: a Croat, a Scot, two Germans (on the same night), a hefty but uninhibited Russian, a Brazilian and a Kurd. “Which was the best?” an interested bystander might ask. “Ah, that would be telling, dear boy, but don’t for a moment suppose I don’t know my way around”. As a well-read closet scientist and Metropolitan wheeler-dealer Huxley would nearly always know what others knew. The question is therefore about how he made use of his knowledge—that is to say, how he mobilized it on the page. His tendency was not to acknowledge other people’s work in any detail except when he wished to rebut it.

Huxley’s alternative mode of operation as a writer was not to make any acknowledgement of other people’s work at all. This was the mode adopted in all the Appleton publications mentioned in this book, which of course in many cases started off as talks and speeches. When he gave a public address, he very rarely mentioned the source of the information being utilized. The point was in the rhetoric, the organization, and the rapprochement with an audience since, while he rarely gave a talk except for a fee, he found the actual performance exhilarating, which is perhaps why he was sometimes reported as changing colour as he spoke. He wanted to impress others as he impressed himself.

An example of how Huxley functioned in this second mode is to be found in his relationship with Henry Maudsley, the distinguished psychiatrist who many years later helped found the Maudsley Hospital in London, an institution particularly important because part of the movement to have the mentally ill treated in a medical facility instead of being incarcerated in a lunatic asylum. That Huxley knew about Maudsley, if he did not know him personally, is certain, since Maudsley's third book, *Responsibility in Mental Disease*, was published in the International Scientific Series in 1874 at exactly the time when Huxley was most involved. Evidently Huxley had not disapproved of this book when he was considering it for publication in the Series. Moreover, the signs are that he took it fairly seriously, learnt from it, and in all probability had a look at the two other books Maudsley had published at that point in his career: *The Physiology and Pathology of Mind* (1867) and *Body and Mind* (1870). What he would have learnt, or what attitude of his own would have been reinforced by Maudsley's experience, was that his ideas about the physical, non-spiritual basis of human existence, were significantly strengthened if mental activity were also taken into account, where the brain and the nervous system somewhat dislodged the concept of mind from any close, medical study of human behaviour. In both their cases, physiological knowledge and understanding were crucial. Huxley had read Maudsley but preferred not to acknowledge the fact. As regards the attention he paid to other people's words and ideas, Huxley could be called the Autolycus of popular semi-academic publishing—a "picker up of inconsidered trifles", like the character with that name in Shakespeare's *A Winter's Tale*. If something useful was at hand, then use it. Autolycus in that play was an enjoyable rogue and petty thief somewhat given to absconding with other people's belongings if they were left around. Likewise, Huxley tended to make use of other people's ideas without acknowledgement when under pressure to write a speech or popular article, though whether this could be branded as outright plagiarism is a moot point. John Lubbock used the word plagiarism when he accused Charles Lyell of using information filched from Lubbock's {…….} in his own book called *The Antiquity of Man*. Lyell expressed surprise. He said he thought it was a writer's job to accommodate the most recent findings and ideas of others. Not without acknowledgement replied John Lubbock.

To determine when plagiarism became a criminal offence rather than a mere folly or oversight is none too easy. Certainly, a student who nowadays uses another person's written or published work without acknowledgement will be penalised and perhaps even dismissed from his or her university without a degree. Among students and professional writers, which is what Huxley was, there can undoubtedly be instances of unconscious borrowing, though this is no longer seen as a defence in either education or the law. The theft of intellectual property is just another form of theft. The Oxford English Dictionary defines plagiarism as "the taking and using as one's own of the thoughts, writings or inventions of another", giving as an example a quotation from the essayist, William Hazlitt: "If an author is once detected in borrowing he will be suspected of plagiarism ever after". More up to date and legally binding definitions have become necessary in the age of electronic technology.

When Huxley gave his three lectures in New York, using what he had learnt at Yale as the main plank of his newly-constructed version of how evolution worked, he mentioned Professor Marsh by name four times, with the implication, however, that while Marsh had the evidence, he, Huxley had the explanation of it. Nothing was said during the course of the three lectures either about the huge research programme in progress in New Haven or about the boldly conceived and implemented acquisition of evidence from the North Dakota Badlands. Marsh's publications on fossil horses were not mentioned either, though, as stated earlier, he had definitely read them. Nor did Huxley comment on or rescind his own earlier publications on fossil horses, where he had taken the horse as an example of a persistent type, one that had remained anatomically constant through long periods of time. He was not the kind of person who would say he had previously been mistaken but had corrected his position on the basis of new evidence. The New York lectures were a performance—an academic ego-trip. When Marsh was asked about Huxley's lectures, he replied ironically that Huxley behaved in that way because he was a great man.

Given the legitimacy of making a public lecture in New York some kind of performance, what then of the publication? Though many lecturers might have envied his ability to carry the lectures off in style, others, especially academic lecturers, would have felt the need for some degree of substantiation, acknowledgement and reference to support what had been said on the day, but Huxley evidently did not have that feeling, which, though, he might have done had he been using his own research results. Nor did Youmans care about such matters. He was simply keen to see the lectures published, not as examples of high-level, well documented scientific research, but as a set of assertions consistent with his own ideas about the need to make science available to a wide audience. He was a popularizer. He knew Huxley was too. The debatable validity of what Huxley said in the third lecture about the evolution of the horse over long periods of time was less important to him than the assertions about the importance of science. He would never know whether what Huxley had said about the anatomy of the horse had been correct or not. Wanting a book that would have a public impact, and sell, he was simply prepared to take Huxley's argument on trust and would never discover whether or not his trust had been justified.

From a professional point of view, Huxley obviously had every right to give direction to his own career more as an advocate for the need for science to be understood and accepted as socially important, than as a research scientist devoted to solving particular problems and having the knowledge, expertise and staying power to do so. When he exercised this right, however,—the right to devote his best energies to the task of influencing what other people thought, with sometimes an almost messianic zeal—there was undoubtedly a price to pay. He never resolved the conflict in his head which bothered him continuously on both the conscious and subconscious levels. This conflict was imbedded in a discrepancy between his studied rationality and belief in logic on the one hand and, on the other hand, the constant threat of irrationality in some mental substratum that was beyond his control. He was seriously ill for almost a month from the day he sailed out of New York Harbour. The emotional and psychological stress compounded during the hectic five weeks in which he had done so much brought him close to breakdown. In her letter to Spencer Baird, thanking him for the photograph that is the frontispiece of the present book, Henrietta would later say: “Mr Huxley was, as you know, very ill both on his voyage home & for some time afterwards; but I rejoice to say, he is now quite well, although often tired from the press of work”.[65] Huxley’s own version was somewhat different. On 31 December 1875 he had told his friend Knowles; “I am subject to Demonical possession and the devil that came to me about ten days ago was a fish”.[66]

Possibly contrary to his expectations, in America Huxley found that his anti-religious diatribes and speeches were not only ineffective, except among those who already lacked religious conviction, but also caused great offence, at least to the extent that people were unimpressed by one type of preaching being replaced by another. Was it really the case that a religious person could not be a scientist? Science might be used to challenge and undermine religious faith, especially if that faith had a simplistic basis in general ignorance of what the physical universe was actually like, but it could not be an alternative to any type of religious conviction, and, as a matter of fact, the more Huxley talked about it, the more this became clear. If it could be said that in America he briefly enjoyed a box-office appeal—a foreigner coming to tell us how to think, while chastising us, explicitly or implicitly for believing in the wrong things—this public appeal ultimately proved unhelpful to Huxley himself, because demonstrating that so many human beings were wrong did not confirm that he was right. When he tried to think about the meaning of his own life, he was left with problems of his own not to be solved by public lecturing.

---

[65] ALS 12 March 1877, Smithsonian incoming correspondence, Vol. 194, pp.541-542.
[66] ALS 31 December 1875, APS B/H981.K

Huxley was constitutionally unwilling to accept that life was meaningless. It had to mean something. Thus like a clever, earnest student thirty years younger than he actually was, a student wishing to comprehend the history of human thought and, within that context, wishing to understand his own actual position as a human being with a mind, Huxley revisited the texts with which he was already familiar, his book about Hume being the immediate, urgent consequence of his worries, at the same time engaging with the work, or rather the publications, of contemporaries like John Stuart Mill, Henry Maudsley and Alexander Bain. Might they have the answers to the problems that deeply troubled him? On the surface of existence he continued to be a formidable public figure who sincerely cared about many of the important issues of the day and was prepared to work towards solutions for the general benefit of society as demonstrated, for example, by his role on the Fisheries Commission and similar public bodies, while in his own home, he continued to be an urbane, witty, caring, stereotypic, Victorian middle-class family man. As far as one can tell, his behaviour both in the work-place and at home was completely sincere, and not merely a façade. On the other hand, his inner—if one can use that word—intellectual and psychological problems remained unresolved. He might appear to be confident, self-assured and in full control of himself in the lecture room and on the printed page, but there was a price to pay for the high level of self-discipline involved. His frequent illnesses, indispositions and breakdowns tend to confirm that this was the case.

The price he paid for full psychological control in public seems to have been his own mental stability or instability, even while he sought convincing intellectual solutions for the problems that worried him. He wanted life to make sense. For example, when he re-read the works of John Stuart Mill, his anxieties about determinism—the determinism that is one of the logical consequences of a belief in science—were somewhat moderated by what Mill said about the two back-alleys or cul-de sacs of any thinking human being: materialism and fatalism. As against these negative, debilitating possibilities, he knew Mill had argued for a degree of personal freedom, for example in the book called *Examination of Sir W. Hamilton's Philosophy*, where Mill says "not only our conduct but our character is in part amenable to our will" and, later, "we can by employing the proper means improve our character". Huxley wanted to believe this. But what were the proper means? Obviously not constantly inveighing against other people's religious beliefs. Mill's escape route from determinism was attractive to Huxley, eventually saving him from the negative or nihilistic social attitudes that were beginning to be prevalent in the theatre, poetry and the arts. The plays of Ibsen that had their first London performances at about this time had, like the poetry of Swinburne, nothing to offer a man who, despite his doubts, still believed it was possible to play a meaningful, constructive role in the world as he happened to find it. John Stuart Mill's works, not least *On Liberty* (1859), *Utilitarianism* (1865) and *Three Essays on Religion* (1874) gave him the landline he needed to escape ship-wreck in the raging storm of the science v. religion conflict. It has to be doubted, however, that he was able to make very much of Mill's book called *The Subjection of Women*. In retrospect, it seems impossible to imagine Huxley changing his view of women. Judging by his attitude to his own daughters, he subscribed to the idea, prevalent at the time, that the physiological differences between men and women simply meant that, while the men did the work, women were designed to stay at home. He knew there were women who wrote books, but they were books for other women and their children. This, bye the bye.

What Mill had to say about character and man's ability to shape his own life, if only to a limited extent, was echoed by at least two contributors to the International Scientific Series: Alexander Bain in *Mind and Body* (1873) and Henry Maudsley in *Responsibility in Mental Disease* (1874). Although Huxley was familiar with both books, and although he knew these two eminent men had made appeal to the will as a vital human resource that would help any human being resist the perils of nihilism, it would have been out of character for him to acknowledge they had assisted him in solving, at least partially, the personal, psychological problems he had carried with him to America. While there he had indulged in rants about religion. Once back home, and revising his lectures, he must have realised that mere rant, however clever, would not do. Perhaps he was not greatly at fault in not acknowledging the influence of Bain and Maudsley, in as far as the concept of will—the ability of the human being to shape his own life intentionally—had a long history, which he knew, though in the light of the mostly negative ideas and attitudes expressed throughout 1876, it might have been more seemly for him to acknowledge that, in reading and re-reading the work of these colleagues, he had been helped towards solutions crucial to the preservation of his own sanity. It was not his way, however, to admit on the printed page that others knew more than he did or had already tried to think their way through problems universally understood to be difficult. He rejected consensus, while, to acknowledge the influence of David Hume, wrote a book about him, gathering the great philosopher into his own lair.

Although a long dissertation might, and perhaps should, be written about the way Huxley sought refuge in the well-established tradition represented by Mill, Bain, Maudsley and others, in effect deviating from his previous course in order to protect his social role, such a discussion would probably be inappropriate in a book about Huxley's visit to America. The point for the moment is to show that when Huxley made such a fuss about other people's beliefs, especially their religious beliefs and the principles by which they conducted their lives, he, as a matter of fact, had not worked out his own position beyond saying repeatedly that he could only accept ideas and opinions supported by the kind of evidence he happened to regard as logical and convincing. Easy to criticize others, but if you wished to take this liberty, you had better be on sound ground yourself. In 1876 Huxley's ideas about science were unsound. Science was not simply a matter of not being religious. Nor could scientific research be reduced to simple formulations easily accessible to the ignorant or uneducated. Boys and girls can look at crayfish and not become scientists. Moreover, true scientists, while always being aware of the work of others, usually conduct their own research, have trustworthy, intelligible methods for gathering and interpreting evidence, and reach conclusions that can be openly discussed, verified and re-checked. Such activity may simply be an end in itself, sometimes proving socially useful, sometimes not; sometimes, perhaps nowadays more frequently, it may be targeted and aimed at the solution of particular problems, but in either case the research needs to be done before conclusions are reached or proposed. We have already seen, at the end of Chapter Two, that Huxley himself admitted this.

Following his trips to France, Germany, Egypt and elsewhere, in search of good health, he had now been to North America, partly for the same reason. Had the strategy worked? It had not. Certainly he had put a brave face on it in his public appearances, on several occasions successfully managing to produce an excellent performance as though doing so came naturally to him, but actually, as regards his health, nothing had been resolved. He was ill on board ship as he went home, Henrietta saying he had caught a chill while on deck, which may of course have been the case. At a deeper level he had expended all the energy he had, so now experienced a reversion, as he said, to being possessed by the old demons that haunted him. There was a great gap between his public performance and his private anxieties; psychologically he had to pay the price of having a professional existence divorced from the normal life he might have had, a condition which certainly has a bearing on the validity of his doctrinaire pronouncements and his rhetorical simplifications. This has never been analysed in a convincing manner, though it easy to see that a psychological price had sometimes to be paid for the Victorian preference, as Huxley adopted it, for keeping up appearances, the effect being that the kind of person one actually was could be effectively disguised. To say this is not a moralistic judgement. On the contrary, many Victorians with psychological problems would have benefited from modern counselling. They had Maudsley's vast experience of mental instability but he would quite often go no further than recommend beef-tea and tennis for his patients. Freud's principle publications began to appear roughly at the time of Huxley's retirement and death. Huxley could not be helped. He was at the mercy, he told Henry James, of an inner torment he could not control. "Phantoms! My dear friend, I live among them—keeping their damned witches' dance at a distance so long as I am fit and well".[67] He had never, though, been able to rely upon being fit and well.

[67] ALS: Huxley to Henry James, 3 November 1894, Houghton Library, Harvard University

APPENDIX

Huxley in the Humboldt Library of Science
(The original title was "The Humboldt Library of Popular Science Literature, so some volumes listed below will have that rubric)

Whereas the present book has been more or less confined to the period 1870 to 1880, and in particular to the year 1876, the Humboldt Library extended over several decades, so has been relegated to an Appendix. Nonetheless, it is reasonably certain that the Humboldt publications, or most of them, had Huxley's blessing since some of them were adapted for this particular publishing programme, while no dispute arose about the legality of the enterprise. Whereas Appleton published his books, the Humboldt Library published some of his articles, so that these, too, were well known in North America during the latter part of the nineteenth century. The Humboldt pamphlets were routinely issued in paper covers, though hard-cover copies could also be obtained at $2 a volume. The publisher was sometimes given as "J. Fitzgerald, New York" and sometimes as "The Humboldt Publishing Company, 64 Fifth Avenue, New York" or "19 Astor Place, New York". A typical subscription was for twelve pamphlets during a twelve-month period. The works were sold only by subscription.

The figures in the left-hand margin of the list that follows are the numbers given to each publication in the series. In Huxley's *Evolution and Ethics* the list has reached number 177. In that volume, the list is preceded by the following statement:

The Humboldt Library of Science is the only publication of its kind, the only one containing *popular scientific works at low prices*. For the most part it contains only *works of acknowledged excellence,* by authors of the first rank in the world of science. In the series are well represented the writings of Darwin, Huxley, Spencer, Tyndall, Proctor, Clifford, Clodd, Bagehot, Bain, Bates,

Wallace, Trench, Romanes, Grant Allen, Geikie, Hinton, Sully, Flammarion, Picton, Williams, Balfour Stewart, Wilson and other leaders of thought in our time. The books are complete and unabridged editions, in neat paper covers.

4 *Evidence as to Man's Place in Nature*
reissues: 1880, 1890 (with three extra essays and a preface written by Huxley)
Locations: LC
Also published, but probably pirated, by Hurst & Co. in New York.

16 *Origin of Species*
This was the ungainly title the publisher or binder in London gave to Huxley's *On Our Knowledge of the Causes of the Phenomena of Organic Nature: Six Lectures for Working Men.*
Locations: LC; Yale; NYPL

21 *The Physical Basis of Life and Lectures on Evolution*
(reprinted from *The Fortnightly Review*)
Locations: APS

36 *Lectures on Evolution: with an Appendix on the Study of Biology*
These are the three lectures delivered in New York, unrevised.
Locations: NLM (1882)

43 *Darwin and Humboldt with Essays by Huxley, Agassiz, Geikie and Thistleton-Dyer*
Locations: NYPL

53 *Animal Automatism, and Other Essays*
The other essays are: "Science and Culture", "On Elementary Instruction in Physiology" (as published in *Nature* 16 (1877): 233-34); "On the Border between the Animal and Vegetable Kingdoms" (as in *Macmillan's Magazine* 33(1876): 373-84) and "Universities: Actual and Ideal"
Locations: LC; NLM; NYPL; JH

66 *Technical Education*
Address on Behalf of the National Association for the Promotion o Technical Education (1887)
Taken from *The Fortnightly Review* 23 (1877): 48-58

96 *A Half-Century of Science*
Edited by Huxley and Grant Allen
Taken from Appleton's *The Advance of Science in the Last Half Century*

114 *Christianity and Agnosticism A Controversy*
A double number incorporating contributions to the *Nineteenth Century*
The Nineteenth Century 25 (1859): 937-64
Huxley's article was from *The Nineteenth Century* 25 (1859): 937-64

125 *Darwinism and Politics* by David Ritchie
incorporating Huxley's "Administrative Realism" from the *Fortnightly Review* 10 n.s. (1871): 525-43

146 *Social Diseases and Worse Remedies*
Locations: NLM (1891)

177 *Evolution and Ethics*
Copies have been seen with the alternative title of
Evolution and Ethics The Romanes Lecture1893
Locations: LC (1894?); US Geological Survey (1896)

HUXLEY'S BOOK PUBLICATIONS

The purpose of the list which follows is to create a complete picture of Huxley's publications in book form, in order to show the flow of work over his life-time. Because of the nineteenth-century habit of putting wrappers round significant, longish articles to circulate among friends and colleagues, it is sometimes difficult to distinguish between printed books and articles of that kind, so this list is merely a first step, intended to be helpful to readers of the present book. The location code is by no means comprehensive. It just indicates where the books listed have been seen by the present author. There will certainly be other libraries with equally important Huxley holdings.

Abbreviations used in the list that follows

AAS American Antiquarian Society, Worcester

APS American Philosophical Society, Philadelphia

BCL Birmingham Central Library

BGS British Geological Survey, Keyworth

CUL Cambridge University Library

F&M Franklin and Marshall University

JH The Johns Hopkins University

LC Library of Congress

LD Leeds Subscription Library

LPN Literary and Philosophical Society, Newcastle

NLM National Library of Medicine, Bethesda

NYPL New York Public Library

MC Michael Collie

RACP Royal Australian College of Physicians

RCS Royal College of Surgeons

RI Royal Institution, London

RS The Royal Society of London

Tex Humanities Research Center, University of Texas at Austin

TRANSLATION: ***A Manual of Human Histology, by Kölliker*** **(1853-4)**

A translation of Rudolph Albert von Kölliker's *Handbuch des gervebehre des menschen* by Huxley and George Busk, in two volumes, for the Sydenham Society

Locations: RCS

On the Educational Value of the Natural History Sciences **(1854)**

(London: John van Voorst, 1854). A pamphlet.

Locations: CUL; RI

The Oceanic Hydrozoa: Descriptions of the Calycophoridae and Pysophoridae,

Ray Society, 1855 (?)

Locations: CUL; Texas; Leeds; BGS; LPN (1859)

TRANSLATION: ***On Tape and Cystic Worms, with an Introduction on the Origin of Intestinal Worms, by Carl Theodor von Siebold*** **(1857)**

Translated for the Sydenham Society

Issued as a two-volume book on intestinal worms and parasites by von Siebold and Kuchenmeister. The first volume was translated by Huxley, the second by Lankester.

British Fossils. Part 1—On the Anatomy and Affinities of the Genus Pterygotus **(1859)**

Memoirs of the Geological Survey of Great Britain, Monograph 1 (1859), pp. 1-36.

In *Scientific Memoirs* (1) Huxley pp.170-202 (2) Salter pp. 203-262, with 16 plates

Locations: BGS

***On Dasyceps Bucklandi (Labyrinthodon Bucklandi, Lloyd)* (1859)**

Memoirs of the Geological Survey (1859), pp. 52-56. In *Scientific Memoirs* Vol pp. 261-68

***Preliminary Essay on the Systematic Arrangement of the Fishes of the Devonian Epoch* (1861)**

Memoirs of the Geological Survey

***On Our Knowledge of the Causes of the Phenomena of Organic Nature (Six Lectures for Working Men* (1863)**

(London: Robert Hardwicke, 1863 and Appleton 1863)

Locations: Hardwicke: NJM (?); RI (1862?); CUL; BCL; Leeds; LPN; NLM; RCS; BGS

In America the title was changed to: ***On the Origin of Species; or, the Causes of the Phenomena of Organic Nature]*** the stereotyped text remaining identical .

Appleton: Texas, 1863 at BCL, APS 1870

The lectures, delivered once a week starting on 10 November 1862, were first published as six separate pamphlets

***Evidence as to Man's Place in Nature* (1863)**

(London: Williams & Norgate, 1863 and Appleton 1863)

Locations: Leeds; MC; APS; RACP (presentation copy); RI; LC; RCS

Appleton locations: JH; F&M (1883?); 1868 (?); posthumous copies at Texas

[Appleton printings listed as 1871, 1872, 1894, 1896 and 1902]

Humboldt (1880?): LC

Hurst & Co: revised text 1890.

***An Elementary Atlas of Comparative Osteology in Twelve Plates* (1864)**

with W.B Hawkins (London: Williams & Norgate, 1864)

Locations: LC; LPN; RCS; BGS

***Lectures on the Elements of Comparative Anatomy; on the Classification of Animals and the Vertebrate Skull* (1864)**

(London: Churchill & Sons, 1864)

Locations: Leeds; RCS

A Catalogue of the Collection of Fossils in the Museum of Practical Geology (1865)

Samuel Laing. *Prehistoric Remains of Caithness, with Notes on the Human Remains* (1866)

(Williams & Norgate, 1866)

pp. [1]-79 by Laing; pp. [81]-148 by Huxley; pp. [149]-161 (the appendix) by Huxley.

Locations: Leeds 1868; LPN; BGS

***Lessons in Elementary Physiology* (1866)**

(London: Macmillan, 1866)

Locations: RCS; RI; RS 1871; 1872; 1881 "new edition"; RI 1885 "revised"; 1893 (revised by Foster)

Translated into German by Rosenthal: NLM

***Elements of Physiology and Hygiene: a Text-Book for Educational Institutions* (1867)**

(New York: Appleton, 1867)

PART I: Elementary Physiology, by Huxley, pp. 11-343

PART II: Elementary Hygiene, by Youmans, pp. 344-452

Questions, Part I, pp. 453-464

Part II, pp. 465-469

Locations: 1867 MC; 1868 NLM; 1872 APS 1873 (revised by Youmans with a new preface); LC 1896

***An Introduction to the Classification of Animals* (1869)**

(London: John Churchill & Sons, 1869)

Locations: RI; LPN; RCS

***Lay Sermons, Addresses and Reviews* (1870)**

(London: Macmillan, 1870)

Locations: RI; Texas; Leeds; RS 1887, 1906; RCS

Appleton: LC 1871

***On the Physical Basis of Life* (1869)**

(New Haven: Charles C. Chatfield, 1869)

The date 1869 is on the titlepage and the introduction is dated 30 March 1869

Locations: NLM; AAS; APS; LC

***Manual of the Anatomy of Vertebrated Animals* (1871)**

(London: J & A Churchill, 1871; New York: Appleton 1872)

Churchill 1871;

Locations: LPN (two copies); RI 1871 & 1877; RCS; CUL; APS (1871: Huxley's personal work copy)

Appleton 1872, 1892

Locations: Appleton 1872 APS

***More Criticisms on Darwin and Administrative Nihilism* (1872)**

(New York: Appleton, 1872)

Locations: APS

This work incorporates reviews of Wallace's *Contribution to the Theories of Natural Selection*, Mivart's *The Genesis of Species*; and Darwin's *The Descent of Man*

Figures and Descriptive Illustrations of British Organic Remains, Decade 13 (1872)

Locations: BGS

***Critiques and Addresses* (1873)**

(London: Macmillan, 1873

Locations: RI;

Appleton 1873: JH

***A Course of Practical Instruction in Elementary Biology* (1876) [check date]**

London: Macmillan 1876

Locations: RCS (1875); JH; RI (1888)

In writing this book, Huxley was assisted by H.N. Martin. The revised edition (Macmillan 1888) was extended and edited by G.B. Howes and D.H.S. Scott, with a new preface by Huxley.

***American Addresses, with a Lecture on the Study of Biology* (1877)**

(New York: Appleton, 1877 and Macmillan 1877)

Macmillan locations: Leeds; RI; RCS

Appleton, JH 1893 (?); NYPL; Academy of Natural Sciences; APS

A Manual of the Anatomy of Invertebrated Animals (1877)

(London: J & A Churchill, 1877)

Locations: RI (2 copies); MC (two copies); RCS

Appleton: APS 1872 (work copy with corrections by Huxley)

Physiography: an Introduction to the Study of Nature **(1877)**

(London: Macmillan, 1877)

Locations: CUL; LPN; BGS

Aberdeen (1878), BGS (1886, revised and cheaper edition); RS (1884. "new edition"), MC (1887); Texas (1893: revised and partly rewritten by R.A. Gregory)

Then: "The lectures on Physiography were delivered at the London Institution in 1869", then "repeated in 1870"

The Crocodilian Remains Found in the Elgin Sandstones, with Remarks on the Ichnites of Cummingstone (1877)

(London: Memoirs of the Geological Survey of the United Kingdom.
Monograph III)

Location: BGS (with the companion volume of plates)

A Catalogue of the Cambrian and Silurian Fossils in the Museum of Practical Geology (1878)

With E.T. Newton

Locations: BGS

Hume **(1878)**

(London: Macmillan 1878) English Men of Letters Series, edited by John Morley

Locations: 1879 (Leeds); RCS (1879);1890 (MC); 1881 (?)

Appleton: 1886 (JH); 1887 (JH)

LC gives Harper [1879]; Lovell [1884]

***The Crayfish* (1879)**

Locations: 1880 (Leeds); MC

American title: ***An Introduction to the Study of Biology, illustrated by the Crayfish***

Locations: APS; JH (1880); RI (1884); MC

Large paper edition with English title: 1880 (F&M #221 of 250)

TRANSLATION: *Freedom in Science Teaching, by Ernst Haeckel*

(Appleton 1879?)

Locations: Leeds

Preface by Huxley

***Introductory Science Primer* (1880)**

(London: Macmillan, 1880)

Locations: RI 1881; Texas 1893

Appleton: LC 1882 with the American title: **Science Primers: Introductory**

***Science and Culture and Other Essays* (1881)**

(London: Macmillan, 1881)

Locations: RS 1881, 1882; RI; CUL; Texas; Leeds

Appleton: NLM (1882)

Fish Diseases (1883)

Inaugural Address at the International Fisheries Congress and Exhibition, volumes 4 and 6

Locations: Birmingham Central; APS

***Animal Automatism and Other Essays* (1884)**

(New York: Appleton, 1884)

Locations: JH; NLM

***The Advance of Science in the Last Half Century* (1887)**

Reprinted from *The Reign of Queen Victoria,* ed. Thomas Humphrey Ward, and the Smithsonian Annual Report for 1887.

Locations: JH; APS; NYPL (1887, 1888, 1889, 1892, 1898)

An Atlas of Practical Elementary Biology

Macmillan 1885

Locations: LPN

***Essays on Some Controverted Questions* (1892)**

(London: Macmillan, 1892 and New York: Appleton 1892)

Macmillan locations: Leeds

Appleton location: JH

***Evolution and Ethics* (1893)**

(Macmillan and Appleton, 1893)

Macmillan: Leeds (1893)

Appleton: JH (1894)

BIBLIOGRAPHY

Althotz, Josef. *Anatomy of a Controversy: the Debate over Essays and Reviews, 1860-1864* (Aldershot: Scolar Press, 1994)

Barr, Alan (ed.) *Thomas Henry Huxley's Place in Science and Letters: Centenary Essays* (Athens GA: University of Georgia Press, 1997)

--------------------- *The Major Prose of Thomas Henry Huxley* (Athens GA: University of Georgia Press, 1997)

Becker, Bernard. *Scientific London* (London: Henry S. King, 1874)

Bibby, Cyril. *T.H. Huxley: Scientist, Humanist, Educator* (London: Watts, 1959)

Block, E. "T.H. Huxley's Rhetoric and the Popularization of Victorian Scientific Ideas, 1854-1874", *Victorian Studies* 29 (1986): 363-86

Clarke, Ronald. *The Huxleys* (New York: McGraw-Hill, 1968)

Clodd, Edward. *Thomas Henry Huxley* (New York: Dodd, Mead & Co., 1902)

Collie, Michael. *Huxley at Work* (London: Macmillan, 1991)

Curle, Richard. *The Ray Society: a Bibliographic History* (London: the Ray Society, 1954)

Dawson, W.R. *The Huxley Papers a Descriptive Catalogue of Papers at Imperial College* (1946)

Desmond, Adrian. *Huxley*
Volume One: *Huxley: the Devil's Disciple* (London: Michael Joseph, 1994)
Volume Two: *Huxley: Evolution's High Priest* (London: Michael Joseph, 1997)

Di Gregorio, M. *Huxley's Place in Natural Science* (New Haven: Yale University Press. 1984)

Exman, Eugene. *The Brothers Harper* (New York: Harper & Row, 1965)

Fiske, John. *Essays Historical and Literary* (London: Macmillan, 1907)

Haeckel, E. "Thomas Henry Huxley", *Nature* 9 (1871): 257-8

Hay, O.P. "Bibliography and Catalogue of the Fossil Vertebrata of North America", *U.S. Geological Survey Bulletin* (1902), 179: 126-7

Howsam, Leslie. *Kegan Paul: a Victorian Imprint* (Toronto: University of Toronto Press, 1998)

Huxley, Leonard. *Life and Letters of Thomas Henry Huxley*, two volumes (London: Macmillan, 1900)

Leidy, Joseph. "Extinct Mammalian Fauna of Dakota and Nebraska", Academy of Natural Sciences of Philadelphia, *Journal* (1869), 2nd. Series, 7: 23-27

Lightman, Bernard. *The Origins of Agnosticism: Victorian Unbelief and the Limits of Knowledge* (Baltimore: the Johns Hopkins Press, 1987)

Lightman, Bernard. *Victorian Popularizers of Science* (Chicago: University of Chicago Press, 2007)

Lubbock, John (Lord Avebury). "Huxley's Life and Work", *Nature* 63 (1900): 92-96. 116-19

Marsh, O.C. "Thomas Henry Huxley", *American Journal of Science* 50 (1995): 177-183

-------------- "Fossil Horses of America", *American Naturalist* 8 (1874): 288-294

Mitchell, Sir Peter Chalmers. *Thomas Henry Huxley: a Sketch of his Life* (New York: G.P. Putnam's Sons, 1901)

Moore, James. *The Post-Darwinian Controversies: a Study of the Protestant Struggle to Come to Terms with Darwin in Great Britain and America* (Cambridge: Cambridge University Press, 1979)

Osborn, H.F. "Biographical Memoir of Joseph Leidy 1823-91", National Academy of Sciences, *Biographical Memoirs* (1913), 7: 335-396

---------------- *Cope, Master Naturalist* (Princeton N.J.: Princeton University Press, 1931)

Paradis, James. *T.H. Huxley: Man's Place in Nature* (Lincoln: University of Nebraska Press, 1978)

Pingree, J. *T.H. Huxley: a List of his Scientific Papers* (London: Imperial College, 1969)

Roos, David. “Neglected Bibliographical Aspects of the Work of T.H. Huxley”, *Journal of the Society of the Bibliography of Natural History*, 8 (1978): 401-20

Schuchert, Charles & C.M. Le Vene. *O.C. Marsh: Pioneer in Palaeontology* (Ne Haven: Yale University Press, 1940)

Shneer, Cecil J. *Two Hundred Years of Geology in America*, Proceedings of the New Hampshire Bicentennial Conference on the History of Geology (Hanover, New Hampshire: the University Press of New England, 1979)

White, Paul. *Thomas Huxley: Making the “Man of Science”* (Cambridge: Cambridge University Press, 2003)

Woolf, Gerard R. *The House of Appleton* (Metuchen NJ: the Scarecrow Press, 1981)

Youmans, E.L. *The Culture Demanded of Modern Life* (

LIBRARY HOLDINGS

Libraries holding Huxley material used in the text

American Antiquarian Society, Worcester
American Philosophical Society, Philadelphia
Birmingham Central Library, England
Cambridge University Library
Franklin and Marshall University
Eisenhower Library, The Johns Hopkins University
Library of Congress
Leeds Subscription Library, England
Literary and Philosophical Society, Newcastle, England
National Library of Medicine, Bethesda, Maryland
Royal Australian College of Physicians
The Royal Institution, London
The Royal Society of London
Humanities Research Center, University of Texas at Austin

Other libraries holding Huxley documents, manuscripts and printed books

University of Oklahoma Library, Norman
Tozzer Library, Harvard
Houghton Library, Harvard
Perkins Library, Duke
Edinburgh University Library
Imperial College, London
Royal College of Surgeons, London
Boston Public Library
The Smithsonian
UCLA
MIT
UNC—Chapel Hill
New York Public Library
British Geological Survey, Keyworth, Nottinghamshire
Lilly Library, University of Indiana
National Library of Scotland
University of Kentucky
MCZ Library, Harvard

INDEX